Basic Concepts: Physical Chemistry Experiments

Neelam seedher, Ph. D
Ex-Professor
Department of Chemistry, Panjab University, Chandigarh, India

Copyright © 2021 Neelam Seedher

All rights reserved

ISBN – 13

Preface

This book is the outcome of my vast experience of teaching physical chemistry to the undergraduate and postgraduate students in the Department of Chemistry, Panjab University, Chandigarh. During the lab sessions, I have observed that there has always been a very serious problem when it comes to understanding the principle of the experiment and the basic concepts involved. This was reflected in the viva-voce tests, conducted from time to time. Most students fared very badly in the viva-voce examination, even after successful completion of the experimental procedure. It had to be emphasized to the students time and again that just learning the solution preparation, operation of the instrument, collecting data, doing calculations and getting the result is not enough. In fact, all this is meaningless unless there is a thorough understanding of the basic concepts involved and the principle of the experiment. Although experimental physical chemistry books usually include principle of the experiment before giving other experimental details, I think most of them miss the required basic concepts. Also it is important to understand that experimental physical chemistry is based on the same fundamental concepts which form the basis of physical chemistry, in general. The present book has therefore been written with the aim of giving students the much needed basic concepts in physical chemistry. To my knowledge this book is the first of its kind, there is no other book available with this kind of content.

In section I, the book gives basic information and general tips about some important aspects of physical chemistry laboratory course. Section II which comprises the main book covers most of the topics included in the undergraduate physical chemistry lab course of various universities. The book has been written in question-answer form so that the students can get precise answers to all the questions which come to their mind related to the understanding of basic concepts in physical chemistry and preparing for the viva-voce tests conducted during the practical examinations of various classes. It may be mentioned that the emphasis is particularly on low-cost experiments and easily available experimental tools in the laboratory.

I hope the students studying chemistry at undergraduate level will find the book very useful. Any suggestions for further improvement of the book will be most welcome.

Neelam Seedher
(Author)

Table of Contents

Preface..iii
Introduction 1
A. Section I (General) 3
1. Introduction to basics 5
2. Accuracy, Precision & Error Analysis 14
3. Units and dimensions 16
4. Cleaning and Calibration of laboratory glassware 26
5. Preparation of solutions 28
6. Graph Plotting 31
7. Safety in physical chemistry laboratory 34
8. Writing a Lab Report 35
B. Section II (Viva-Voce Questions) 39
1. General 41
2. Accuracy, Precision and Error Analysis 52
3. Units and dimensions 66
4. Preparation of solutions 72
5. Distribution law 80
6. Phase rule 88
7. Thermochemistry 97
8. Acid-base indicators 110
9. pH metry 117
10. Colorimetry 130
11. Liquid state – Density 146
12. Liquid state - Surface tension 150
13. Liquid state - Refractive index 163

14. Liquid state - Viscosity 172
15. Solutions – Partial molar volume 184
16. Adsorption 190
17. Colloids 199
18. Potentiometry 209
19. Chemical kinetics 231
20. Cryoscopy 244
21. Conductometry 251
22. Polarimetry 270
23. Dielectric constant and Dipole moment 279
General Bibliography 293

Introduction

Physical chemistry is an experimental science. Physical chemistry laboratory exposes students to fundamental concepts of physical chemistry using some modern experimental equipment and techniques. This laboratory also teaches them how to develop the ability to use standard mathematical analyses to correctly describe the numerical significance of experimental results as well as the ability to develop scientific communication skills through oral quizzes, written reports and presentations. Learning procedures for carrying out experiments, the solution preparation, operation of the instrument, collecting data, doing calculations and getting the result are some of the steps involved while undertaking a course in experimental physical chemistry. However, all this is meaningless unless there is a thorough understanding of the basic concepts involved and the principle of the experiment. Moreover, it is important to understand that experimental physical chemistry is also based on the same fundamental concepts which form the basis of physical chemistry, in general. The present book written in question-answer form will be a useful reference text for getting answers to a wide range of questions which come to mind related to the understanding of basic concepts and preparing for the viva-voce tests.

Section I includes introduction to basics and some general aspects such as accuracy, precision and error analysis, units and dimensions, unit conversions, cleaning and calibration of glassware, preparation of solutions, plotting a graph by hand, safety in physical chemistry laboratory and writing a lab report. Section II is in the question answer format and deals with various general topics as well as some specific topics related to the experiments carried out in undergraduate physical chemistry laboratories.

A. Section I (General)

1. Introduction to basics

Almost all experimental work in physical chemistry laboratory involves measurement of weight and volume as well as the measurement and control of temperature. Density of liquids is another property used in many experiments. It is therefore, essential to discuss these aspects in some detail.

i)   **Measurement of weight**

Precision weighing is a necessity in any physical chemistry laboratory. A weighing balance or weighing scale is a devise used to measure mass or weight. In scientific work for accurate measurement of very small weights analytical balances are used. Although a wide range of such devices are in use, laboratory balances can be broadly classified into following types.

Mechanical balance: In mechanical balances, there is no power requirement. They are generally of two types: pan or beam balance and spring balance. A beam balance consists of a beam supported at the center by an agate knife edge resting on a support moving inside a vertical pillar. It works on the principle that when two bodies with equal masses are placed on the pans they secure a state of horizontal balance of the beam. A spring balance consists of a spring fixed at one end with a hook to attach an object at the other. Greater the load more is the stretching. It works by Hooke's Law which states that the stress on the spring is proportional to the strain. In other words, the force needed to extend a spring is proportional to the distance that spring is extended from its rest position.

An important difference between a beam (pan) balance and a spring balance is that a beam balance measures the mass of an object while a spring balance measures weight which is mass multiplied by the acceleration due to gravity. Weight is a measurement of how much the force of gravity acts on a given amount of mass. So the weight of an object changes with change in the acceleration due to gravity while the mass remains unaffected. Therefore, beam balances are used for the precision measurement of mass as their accuracy is not affected by variations in the local gravitational field. A change in the strength of the gravitational field caused by moving the balance does not change the measured mass, because the moments of force on either side of the beam are affected equally. Thus a beam balance gives an accurate measurement of mass at any location experiencing a constant gravitational force. However, in general use, weights expressed in mass units (such as kilograms), are accepted because any mass on earth converts to a weight in pretty much the same way and the acceleration due to gravity varies only slightly all over earth. But scientifically speaking, weight is a force and should be expressed in newton, the unit of force.

Digital electronic balances: The beam balances and spring scales have now been replaced by digital electronic balances in most physical chemistry labs. Electronic balances offer digital display and are amongst the most precise measuring devices designed to measure small mass in sub milligram range. Electronic scales are compact consisting of a single pan and an electronic sensor to display the weight of the specimen. Since they are sensitive to environmental factors, the measuring pan is placed inside a transparent enclosure, called draft shield which has doors so that dust particles and any air currents in the room do not affect the operation of the balance. The sample must also be at room temperature to prevent natural convection from forming air currents inside the enclosure and causing an error in reading.

Electronic analytical scales measure the force needed to counter the mass being measured rather than measuring the actual masses. As such they must have calibration adjustments made to compensate for local gravitational field. Modern electronic laboratory balances work on the principle of electromagnetic force restoration. Such systems use an electromagnet to generate a force to counter the sample being measured. The force imposed on the system by the weight is compensated by the current through the coil. A detector measures the current required to oppose the downward motion of the weight in the magnetic field.

Checking accuracy: Accuracy checks of a weighing balance involve testing for reproducibility, linearity and calibration. Reproducibility refers to the instrument's ability to repeatedly deliver the same weight reading for a given object. Linearity is the characteristic which quantifies the accuracy of the instrument at intermediate readings throughout the weighing range of the instrument. Calibration refers to a comparison of the weight reading of a given mass standard, and the actual value of that standard.

Precautions: In order to avoid weighing errors while using an analytical electronic balance, the following points must be taken care of. Weighing errors can be internally or externally induced. Internally induced errors may stem from improper handling of the balance, overloading it or otherwise damaging the delicate weighing mechanism. In such a case the balance should be returned to the manufacturer for repair by trained technicians. Externally induced errors are easy to reduce or eliminate by taking some necessary precautions.

Environmental conditions for best weighing accuracy: i) Balance should be kept on a vibration-free support. ii) Always check the leveling of the balance with the help of leveling bubble provided. iii) Dust and air currents greatly affect the accuracy of measurements. Close doors of the balance enclosure before weighing. Also the balance should not be located next to doors or windows as opening and closing gives rise to air drafts. iv) It is also recommended to keep the balance in an area with controlled temperature. The sample temperature and balance temperature must be the same. Equilibrate hot or cold samples to room temperature in closed containers prior to weighing. When the sample temperature is higher than the room temperature, a layer of warmer air around the sample creates a slight upward air current which causes error. Place the balance away from temperature extremes such as direct sunlight, heating and air conditioner vents. v) Silica gel bag should be placed inside weighing chamber to keep the chamber free from humidity. The bags should be periodically recharged or replaced. vi) Sample weight can change due to absorption of or giving off moisture or other volatile components. In such cases, the sample weight does not stabilize but gradually continues to increase or decrease. Protect hygroscopic samples from moisture until just prior to weighing. vii) Keep the inside of the weighing chamber and surrounding bench space scrupulously clean so as to avoid errors and cross contamination of samples. viii) Every time use a clean spatula of appropriate size. ix) It is advisable to tare the balance after placing the weighing container so that its reading is zero. This means that the final reading will be that of the material to be weighed and will not reflect the weight of the container. Most balances allow taring to 100% of capacity. x) Avoid overloading of the balance.

(Further reading: Laboratory Balances: How they work, checking their accuracy, Douglas Morse BS, Daniel M and Baer MD. https://academic.oup.com/labmed/article-pdf/35/1/48/24958905/labmed35-0048.pdf)

ii) Measurement of volume

Almost all physical chemistry experiments require precise measurement of the volume of liquids. The primary requirement for this purpose is the use of thoroughly clean and properly calibrated volumetric glassware. The cleaning and calibration of glassware is discussed in detail separately in a subsequent section.

The various kinds of volume measuring tools used for this purpose vary in the degree of accuracy. For example, the accuracy of graduated beaker and measuring cylinder is low as compared to other tools such as volumetric flask, burette and pipette. Therefore, graduated beaker or measuring cylinder cannot be used for preparation of standard solutions. For dispensing very small volume of liquids (0.1–1000 µL) precisely, micropipettes should be employed. After adjusting the volume to be dispensed, a plunger is depressed by the thumb and as it is released, liquid is drawn into a disposable plastic tip. When the plunger is pressed again, the liquid is dispensed. Precise volumes must always be measured at constant temperature because due to thermal expansion, changes in temperature are followed by changes in volume. Volumetric vessels are usually calibrated by manufacturers at 20°C.

iii) Measurement and control of temperature

Temperature is another basic parameter which greatly affects almost all physico-chemical processes. Physical chemistry labs are therefore, always equipped with appropriate devises for the measurement as well as control of temperature.

Measurement of temperature: In ordinary laboratory work usually the following types of devises are used for measurement of temperature.

a) Liquid-in-glass thermometers: Although any liquid with a thermal expansion coefficient significantly greater than that of glass can be used, the thermometers in common usage are mercury and alcohol thermometers.

Mercury thermometer, invented by physicist Daniel Gabriel Fahrenheit in 1714, is the most commonly used thermometer in lab. Mercury has a number of advantages over other liquids for use in thermometers. For example, mercury does not wet glass, remains in liquid form over a wide range of temperature (- 37 to 356^0 C), has high thermal conductivity and can be purified easily. The major disadvantage of mercury thermometer is the health hazards posed by mercury, if the thermometer breaks accidently. Due to this reason mercury is sometimes replaced by alcohol. However, alcohol thermometers show less accurate reading and are slower in response as compared to mercury thermometers. This is because alcohol wets glass surface, has lower thermal conductivity and smaller temperature range.

Mercury thermometer works on the principle that mercury expands on heating. It basically consists of a glass bulb for mercury storage which is attached to the stem. The stem consists of a narrow capillary tube of uniform bore enclosed in a glass tube. A scale is scribed on the outside of the glass tube and it is calibrated so that the height to which mercury rises in the tube directly gives the temperature reading. To construct mercury thermometer, a clean fine glass capillary tube having a bulb at one end is taken. The glass bulb is filled with pure and dry mercury by alternate heating and cooling. The open end of the capillary tube is then vacuum-sealed. Mercury expands when the temperature rises. A small change in the volume of mercury drives the narrow mercury column a relatively long way up the tube. In order to calibrate the thermometer, two fixed points of temperature are chosen. The bulb is made to reach in thermal equilibrium with ice-water mixture to mark the lower fixed temperature and boiling water in equilibrium with its vapour to mark the upper fixed temperature. The thermometer is graduated in equal intervals between these two fixed marks. Common thermometers are graduated to 0.1^0. For precise work, thermometers with graduation of 0.01 or 0.02^0 are also available.

b) Thermocouples: A thermocouple is a sensor which is used to measure temperature. The thermocouple measures unknown temperature of the body with reference to the known temperature of the other body. It consists of two metals joined together to form two junctions. One of the junctions is connected to the body whose temperature is to be measured (hot or measuring junction) while the other junction is connected to a body of known temperature (cold or reference junction). Since the two junctions are maintained at different temperatures the Peltier emf is generated within the circuit and it is a function of the temperatures of two junctions. The total emf generated in the circuit depends on the metals used in the circuit and the temperatures of the two junctions. If both the junctions are at the same temperature, equal and opposite electromotive force (emf) will be generated at both junctions and the net current flowing through the junction is zero. If the two junctions are at different temperatures, the emf will not become zero and there will be a net current flowing through the circuit. The total emf or the current flowing through the circuit can be measured easily by a suitable device. Thermocouples can measure temperatures over a wide range from - 250^0C to 1700^0C.

c) Thermistors: The term thermistor is a combination of "thermal" and "resistor". Thermistors are thermally sensitive resistors which exhibit a large, predictable and precise change in electrical resistance when subjected to change in temperature. Thermistors consist of small beads of a mixed metallic oxide semiconductor to which are attached platinum leads; the beads about one mm in diameter are encapsulated in a glass tube. There are two types of thermistors: Negative Temperature Coefficient (NTC) and Positive Temperature Coefficient (PTC). In an NTC thermistor, there is inverse relationship between resistance and temperature, increase in temperature decreases resistance and vice versa. This type of thermistor is very commonly used. The working of a PTC thermistor is different. When temperature increases, the resistance increases and when temperature decreases, resistance decreases. This type of thermistor is generally used as a fuse. Because of their very predictable characteristics and their excellent long term stability, thermistors are generally accepted to be the most advantageous sensor for many applications including temperature measurement and control.

d) Digital thermometers: Digital thermometers offer a number of advantages over conventional mercury thermometers. One major advantage of digital thermometers is that they are unbreakable and mercury free. Mercury thermometers have to be handled very carefully due to the hazards of mercury spillage on accidental breakage. Moreover, digital thermometers are fast, easy to read and convenient to handle as compared to the conventional thermometers. They are based on either thermistor or thermocouple sensors and contain a small computing mechanism. The computer converts the output signal into temperature and offers a digital readout in degrees. The temperature can be read exactly and accurately since the temperature reading does not depend on scale reading and instead is shown directly on the display. Also digital thermometers can reach final temperature in lesser time (5 to 10 seconds) compared to conventional thermometers.

Temperature control: Temperature control is an important aspect since almost all physico-chemical properties are affected by temperature. Many experiments need to be performed at constant temperature. Constant temperature baths or thermostats are therefore an essential part of equipment in any physical chemistry laboratory. A thermostat is a device which switches the heating/cooling system on and off as necessary to maintain constant temperature. The thermo regulator is the most important component of a thermostat. Earlier toluene-bulb regulators or mercury-in-glass regulators were used but now they have been largely replaced by other sensor types. Common sensor technologies in use today include bimetallic mechanical or electrical sensors, electronic thermistors, semiconductor devices and electrical thermocouples.

Laboratory water bath is the most commonly used type of thermostat in physical chemistry labs. Water bath consists of a container called bath, of appropriate size made up of glass or metal with glass windows containing water. A circulating pump is used to circulate water. It is equipped with a heating/cooling mechanism consisting of a thermo regulator and a relay which makes or breaks the electric circuit. Water can be replaced by oil as thermostating liquid when temperatures higher than 100^0C are desired which is rare in physical chemistry labs. Circulating water baths provide more uniform temperature and are thus more accurate than non-circulating water baths. Shaking water baths are also available for some specific purposes.

iv) Density of liquids

Density is the amount of matter in a given volume. Density, also called mass density of a liquid, is defined as the mass of unit volume of the liquid. It is measured in g cm^{-3} in CGS units and kg m^{-3} in SI units. Density and specific gravity are not identical measures. Specific gravity is the relative density of a material. It is measured relative to the density of a reference substance. Since the reference substance is usually taken as water, the ratio of the density of given liquid to that of water is defined as the specific gravity of the liquid. It is also referred to as the relative density with respect to water. Density is an absolute quantity whereas specific gravity is a relative quantity with no units. Specific gravity values are useful only for predicting whether or not a given liquid will float on water and for comparing the densities of different liquids. Another related term, bulk density is used primarily for materials consisting of solid particles and is defined as the mass of many particles of the material divided by the total volume they occupy. The total volume includes particle volume, inter-particle void volume, and internal pore volume. Similarly number density, defined as the number of particles per unit volume, is used mainly for gases. The usefulness of density is that it provides a link between mass and volume. Mass and volume are extensive properties whereas density is an intensive property.

Measurement of density: The density of a liquid can be measured by hydrometer based on the relation of buoyancy to density. Relative density is also determined by weighing a definite volume of the liquid in a specific gravity bottle. The specific gravity bottle is replaced by pycnometer for greater accuracy. Pycnometer is weighed empty, full of a liquid with a known density (normally water) and full of the liquid whose density is to be determined. The Pycnometer must be thermostated precisely to the reference temperature since density is temperature-dependent. For a fixed volume of the liquid, relative density or specific gravity is given by the ratio of the mass of sample to the mass of water. The density value can be determined by multiplying the relative density by the density of water at that temperature. These are simple, relatively inexpensive methods but require a large sample volume and are not very precise. Hydrostatic balances which determine the buoyancy acting on a sinker immersed in the liquid (Archimedes principle) are more reliable and precise but are time consuming, expensive and difficult to set up. Digital density meters which are much faster, convenient and precise but expensive are now being increasingly used where high precision is required.

2. Accuracy, Precision & Error Analysis

Accuracy and precision are the most important concepts for any experimental measurement. The two terms cannot be used interchangeably since each conveys a different meaning in scientific measurements. Accuracy refers to the agreement between experimental data and a known value while precision refers to how well experimental values agree with each other. In other words, accuracy defines the closeness of a measured value to a standard or known value while precision expresses the degree of reproducibility or agreement between repeated measurements. A data can be very precise such that each data point is close to the others but least accurate if the measured value is far from the actual or known value.

All experimental measurements have a degree of uncertainty due to errors in the measured result. Error in the measurement of a physical quantity is defined as its deviation from actual value. It is important to understand the type and source of such errors. Usually the errors are either due to the limits of the measuring instrument or the skill of the experimenter making the measurements. Accordingly errors are classified as systematic errors, random errors and blunders. Systematic errors are associated with some fault either in the measuring devise or the design of the experiment. For example, there may be problem with the calibration of the instrument or its data handling system. Systematic errors tend to be consistent in magnitude and/or direction and cannot be reduced by repeated measurements using same instrument and the same methodology. These errors are difficult to detect and cannot be analyzed statistically. Systematic errors can only be identified and eliminated after careful inspection of the experimental methods used, cross-calibration of instruments and examination of techniques employed.

Random errors are unpredictable random fluctuations in the measured value. Repeated measurements yield results that fluctuate above and below the true or accepted value and thus they affect the precision of a measurement. Possible sources of random errors can be i) observational; for example, the experimenter's inability to get the same value when reading the scale of a measuring device to the smallest division, ii) environmental; for example, unknown and unpredictable voltage or temperature fluctuations or mechanical vibrations of equipment iii) limitations in the sensitivity of measuring instruments. Since such errors occur in either direction, they can be minimized by taking average of a large number of readings and can often be evaluated by statistical analysis. That is why they are also called statistical errors.

Blunders, also called personal errors are simply mistakes, for example in reading a scale, recording data, doing calculation etc. These can usually be identified and corrected by careful re-examination and repetition of the experimental procedure and calculations.

Error analysis is the study of uncertainties in physical measurements. Statistical analysis of a measurement repeated several times allows us not only to obtain a better idea of the actual value, but also enables us to characterize the uncertainty of the measurement. The two important parameters in the statistical analysis; the mean which gives the central or average value and the standard deviation which describes the spread or deviation of the measured values about the mean, can be used to determine the uncertainty of measurement.

(Further reading: Measurements and Error Analysis. College Physics Labs, Mechanics, The university of North Carolina at Chappal Hill. Web site: http://www.webassign.net/question_assets/unccolphysmech11/measurements/manual.html)

3. Units and dimensions

There is a difference between units and dimensions. The dimensions of a quantity refer to the basic nature of the quantity without numerical value while unit is a way of measurement of that dimension. For example, length is a dimension but it is measured in units of feet or meter. A given quantity can be reported in many different kinds of units, but it will always have the same dimensions. For example, force which is equal to mass x acceleration, always has dimensions of (mass x length)/time^2, but it can be expressed in different units such as Newton, dyne etc.

Unit Systems: Any physical quantity is expressed as the product of a numerical value and a unit. Just as a person is known by his or her name, the physical quantities are known by the units of measurement. There are three common categories of unit systems: CGS (centimeter, gram, second) system, MKS (meter, kilogram, second) system and SI (International system of units) system. CGS system, used in earlier days, has now been largely replaced by MKS system. The MKS system differs from CGS system only in the scale of two of the three basic units of length and mass. The International System of Units (abbreviated SI from systeme internationale, the French version of the name) is the current globally agreed system of units. The main difference from MKS system is that while MKS system has only three base units (mass, kilogram, second), SI system includes four other base units, namely ampere, kelvin, mole and candela, that measure electrical current, temperature, amount of substance and luminous intensity, respectively. All other SI units can be derived from these seven base units.

SI system of units: All systems of weights and measures are linked through a network of international agreements supporting the International System of Units. The International System is called the SI, using the first two initials of its French name *Système International d'Unités*. The SI is maintained by the International Bureau of Weights and Measures (BIPM, for *Bureau International des Poids et Mesures*), and it is updated every few years by an international conference, the General Conference on Weights and Measures (CGPM, for *Conférence Générale des Poids et Mesures*), attended by representatives of all the industrial countries and international scientific and engineering organizations. There are two classes of units in the SI system: base units and derived units.

SI Base Units: The seven base units from which all other SI units of measurement are derived are included in Table 1 along with their names and symbols. The base units relate to seven fundamental scientific quantities. The other physical quantities can be derived from these units. The names of all SI units begin with a lower case letter except at the beginning of a sentence. SI symbols are always written in lower case except for the liter and those units derived from the name of a person (e.g., A for ampere, named after French mathematician and physicist André-Marié Ampere; K for kelvin, named after the Belfast-born, Glasgow University engineer and physicist William Thomson, 1st Baron Kelvin etc.). SI symbols should not be followed by a period (except at the end of a sentence). Likewise, symbols stand for both the singular and plural of the unit and should not have an "s" added when more than one.

Table 1. The seven base SI units

Physical Quantity	Unit Name	Unit Symbol
Mass	kilogram	kg
Length	meter	m
Time	second	s
Temperature	Kelvin	K
Amount of substance	mole	mol
Current	Ampere	A
Luminous Intensity	candela	cd

Supplementary SI units: Two supplementary SI units are radian (rad) for plane angle and steradian (sr) for solid angle.

SI unit definitions:

i) kilogram (kg): The kilogram is the mass of the platinum-iridium prototype which was approved by the Conférence Générale des Poids et Mesures, held in Paris in 1889, and kept by the Bureau International des Poids et Mesures.

ii) meter (m): The meter is the length of the path travelled by light in vacuum during a time interval of 1/299 792 458 of a second.

iii) second (s): The second is the duration of 9 192 631 770 periods of the radiation corresponding to the transition between the two hyperfine levels of the ground state of the cesium 133 atom.

iv) Kelvin (K): The kelvin is the fraction 1/273.16 (3.6609×10^{-3}) of the thermodynamic temperature of the triple point of water.

v) mole (mol): The mole is the amount of substance which contains as many elementary entities as there are atoms in 0.012 kilogram of carbon 12.

vi) Ampere (A): The ampere is the intensity of a constant current which, if maintained in two straight parallel conductors of infinite length, of negligible circular cross-section, and placed 1 meter apart in vacuum, would produce between these conductors a force equal to 2×10^{-7} newton per meter of length.

vii) candela (cd): The candela is the luminous intensity, in a given direction, of a source that emits monochromatic radiation of frequency 540×10^{12} hertz and that has a radiant intensity in that direction of 1/683 watt per steradian.

A radian cuts out a length of a circle's circumference equal to radius while a steradian cuts out an area of a sphere equal to $(radius)^2$.

SI Derived Units: Other SI units are called SI derived units. There are numerous SI derived units which are complementary to the base units. Some commonly used SI derived units, expressed in terms of SI base units are given in Table 2. Certain SI derived units have been given special names and symbols, and these special names and symbols may themselves be used in combination with the SI and other derived units to express the units of other quantities. For example, the SI derived unit of momentum (mass times velocity) has no special name; momentum is stated in kilogram meters per second ($kg\ m\ s^{-1}$) whereas the SI derived unit of force is given a special name, Newton (symbol, N). Some common SI units with special names are given in Table 3.

Certain non-SI units such as minute (symbol, min = 60 s), hour (symbol, h = 60 min = 3600 s), liter (symbol, L = 1 dm^3 = 10^3 cm^3) are usually accepted in the SI system due to their everyday use. However, some other

Table 2. Some commonly used SI derived units, expressed in terms of SI base units.

Quantity	Name	Symbol
Acceleration	meter per second squared	m s^{-2}
Angular momentum	kilogram meter squared per second	kg m^2 s^{-1}
Area	meter squared	m^2
Concentration	mole per cubic meter	mol m^{-3}
Density	kilogram per cubic meter	kg m^{-3}
Diffusion coefficient	meter squared per second	m^2 s^{-1}
Electric current density	ampere per square meter	A m^{-2}
Magnetic field strength	ampere per meter	A m^{-1}
Magnetic moment	ampere meter squared	A m^2
Mass per unit area	kilogram per square meter	kg m^{-2}
Mass per unit length	kilogram per meter	kg m^{-1}
Molality	mole per kilogram	mol kg^{-1}
Molar mass	kilogram per mole	kg mol^{-1}
Molar volume	cubic meter per mole	m^3 mol^{-1}
Momentum	kilogram meter per second	kg m s^{-1}
Specific volume	cubic meter per kg	m^3 kg^{-1}
Speed	meter per second	m s^{-1}
Velocity	meter per second	m s^{-1}
Volume	cubic meter	m^3

Table 3. SI derived units with special names

SI derived quantity	Name	Symbol	Equivalent other SI units	Equivalent SI base units
Conductance	Siemens	S	Ω^{-1}	$kg^{-1} m^{-2} s^3 A^2$
Electrical resistance	ohm	Ω	VA^{-1}	$kg\, m^2 s^{-3} A^{-2}$
Electric potential	volt	V	WA^{-1}	$kg\, m^2 s^{-3} A^{-1}$
Electric potential	volt	V	JC^{-1}	$kg\, m^2 s^{-3} A^{-1}$
Energy	joule	J	Nm	$kg\, m^2 s^{-2}$
Force	newton	N	$kg\, m\, s^{-2}$	$kg\, m\, s^{-2}$
Frequency	hertz	Hz	s^{-1}	s^{-1}
Power	watt	W	Js^{-1}	$kg\, m^2 s^{-3}$
Pressure	pascal	Pa	Nm^{-2}	$kg\, m^{-1} s^{-2}$
Quantity of electricity	coulomb	C	As	As
Work	joule	J	Nm	$kg\, m^2 s^{-2}$

units which are the older versions of metric system such as calorie, micron, millimicron, mho etc. are not recommended for use in the SI system of units.

SI Prefixes: Prefixes are used to avoid very large and very small numerical values of physical quantities. The SI prefixes are standardized for use in the International System of Units (SI) by the International Bureau of Weights and Measures (BIPM). For example, the prefix, kilo in kilometer indicates that the distance is 1000 times larger than the meter. Similarly for measuring short lengths such as 1/1000th of a meter, millimeter is used. Each prefix has a unique symbol. SI system defines twenty prefixes varying from a factor of 10^{24} to 10^{-24}. The commonly used SI prefixes, however have multiplication factor varying from 10^{12} to 10^{-12} and are given in Table 4. Prefixes may not be used in combination. For example, the SI base unit of mass, kilogram already contains a prefix, kilo and therefore, milligram (mg) is used instead of microkilogram (μkg) for 10^{-3} gram.

Table 4. The prefixes used in the International System of Units (SI)

Factor	Name	Symbol	Numerical value
10^{12}	tera	T	1 000 000 000 000
10^{9}	giga	G	1 000 000 000
10^{6}	mega	M	1 000 000
10^{3}	kilo	k	1 000
10^{2}	hecto	h	100
10^{1}	deka	da	10
10^{-1}	deci	d	0.1
10^{-2}	centi	c	0.01
10^{-3}	milli	m	0.001
10^{-6}	micro	µ	0.000 001
10^{-9}	nano	n	0.000 000 001
10^{-12}	pico	p	0.000 000 000 001

Some fundamental physical constants: Some quantities, called the fundamental physical constants with specific and universally used symbols, are of such importance that they must be known to high degree of accuracy. Table 5 provides the values in SI units for some commonly used physical constants.

Table 5. Values in SI units for some common physical constants*

Constant	Symbol	Value**
Acceleration due to gravity	g	9.806 m s^{-2}
Atomic mass unit	amu	1.660 x 10^{-27} kg
Avogadro's No.	N	6.022 x 10^{23} mol^{-1}
Boltzmann constant	k	1.381 x 10^{-23} J K^{-1}
Coulomb constant	$1/4\pi\varepsilon_0$	8.987 x 10^{9} N m^2 C^{-2}
Electron charge	e	1.602 x 10^{-19} C
Electron rest mass	m_e	9.109 x 10^{-31} kg
Faraday constant	F	9.649 x 10^{4} Cmol^{-1}
Gas constant	R	8.314 J mol^{-1} K^{-1}
Neutron rest mass	m_n	1.675 x 10^{-27} kg

Constant	Symbol	Value**
Permeability of vacuum	μ_0	$4\pi \times 10^{-7}$ N A^{-2}
Permittivity of vacuum	ϵ_0	8.854×10^{-12} F m^{-1}
Planck constant	h	6.626×10^{-34} J s
Proton rest mass	m_p	1.673×10^{-27} kg
Speed of light in vacuum	C	2.998×10^8 m/s
Standard atmosphere	atm	101.325 kPa

*Fundamental Physical Constants-Hyperphysics. Ref.: CODATA, Internationally recommended 2014 values of fundamental physical constants, Physical reference data, Physical Measurements Laboratory, National Institute of Standards and Technology (NIST).
** The reported values have been rounded off to three decimal places.

Unit conversions: Most physical quantities can be expressed in more than one unit of measurement and conversion from one unit to another is often necessary. Conversion of units usually involves multiplication or division by an appropriate conversion factor. It is also important to be careful about the selection of the correct number of significant digits and the rounding strategy for the figure in the converted unit. The rules for counting the number of significant figures are the same as applicable for other calculations. These rules are discussed in detail in section II. For multiplication and division, the answer should have the same number of significant figures as the term with the fewest number of significant figures while for addition and subtraction, round off to the number of decimal places as the term with the fewest decimal places. In the commonly used dimensional analysis method, units are treated like numbers and can thus be multiplied and divided (though not added or subtracted) just as numbers can. We have to set up an equation so that unwanted units cancel, leaving only the desired units. For example, an athlete covers a distance of 400 m in 45.18 s. What is the average speed of the athlete? Average speed = 400 m/45.18 s = 8.853474989 m/s. The calculated speed cannot be accurate up to 9 decimal places. For rounding this figure we apply the concept of significant figures. Out of the two values going into the calculation, 400 m is an exact number and exact numbers are treated as having infinite number of significant figures while time 45.18 s has four significant figures. Therefore, the answer should be rounded off to four significant figures and the speed should be reported as 8.853 m/s. Similarly speed in km/h calculated by dividing with 1000 and multiplying with 3600 (the number of seconds in 1 hour) should be reported as 31.87. Unit conversion can be from one SI unit to another or from a non-SI unit to SI unit. Many non-SI units are now defined exactly in terms of SI units while others are related to SI units via fundamental constants. Since some non-SI units are also frequently used in the science literature, it is usually desired to convert all non-SI units and prefixed SI units to base SI units before carrying out calculations. Non-SI to SI conversion factors are given in Table 6 for some commonly used units.

Table 6. Non-SI to SI conversion factors for some commonly used units*.

Non-SI unit	Symbol	SI equivalent
ångstrom	Å	10^{-10} m
atmosphere	atm	101325 Pa
atomic mass unit	u	1.661×10^{27} kg
bar	bar	10^5 Pa
calorie	cal	4.184 J
debye	D	3.336×10^{-30} Cm
electronvolt	eV	1.602×10^{-19} J
hour	h	3600 s
litre	l, L	10^{-3} m^3
micron	μ	10^{-6} m
minute	min	60 s
degree Celsius	^{0}C	^{0}C + 273.15 K
degree Celsius (temp. interval)	^{0}C	1 K**

Non-SI unit	Symbol	SI equivalent
degree Fahrenheit	K	$°F + 459.67/1.8$ K
degree Fahrenheit (temp. interval)	°F	5/9 (=0.5556) K***
kilowatt hour	kW h	3.6×10^6 J
metric ton	t	10^3 kg

* Values taken from NIST Guide to the SI, Appendix B: Conversion factors, Physical measurements laboratory, National institute of standards and technology (NIST).

** A temperature difference of $1°$ C is equivalent to a temperature difference of 1K, meaning the unit size in each scale is the same. Only the scales start in different places. Zero on the Celsius scale (0°C) is defined as equivalent to 273.15K and therefore, K = °C +273.15.

*** A temperature difference of 1°F is the equivalent of a temperature difference of 5/9 = 0.556°C. A temperature difference of 1°C is the equivalent of a temperature difference 1.8°F. Fahrenheit is a thermodynamic temperature scale, where the freezing point of water is 32°F and the boiling point 212°F (at standard atmospheric pressure). °C = (°F – 32)/1.8.

Chemical Conversions: Chemical conversions are the conversions between different units of chemical quantities such as grams, moles, number of molecules, number of atoms etc. Mole, the unit of measurement for the amount of substance, is the central unit in chemical conversions. Mole is defined as a number which represents the number of atoms in 12 grams of the carbon-12 isotope. This number is equal to 6.023×10^{23} and is called Avogadro's number. However, mole does not necessarily represent the number of atoms; we can have a mole of anything like a mole of particles, a mole of molecules, a mole of ions etc. For example, the number of molecules in 10 moles of water is $10 \times 6.023 \times 10^{23} = 6.023 \times 10^{24}$.

Similarly 6×10^{10} molecules of sodium hydroxide are approximately equal to 10^{-13} moles of sodium hydroxide. The molar mass, the sum of the atomic masses of the atoms making up the molecule, is used to convert grams of a substance to moles. Atomic mass is the mass in grams of 1 mole, 6.022×10^{23} atoms, of an element. The mass of a mole of substance is called the molar mass of that substance. Weight in grams of a substance divided by the molar mass in grams is equal to the number of moles. For example, one mole of carbon dioxide weighs 44 grams and 22 grams of carbon dioxide is equal to 0.5 mole of carbon dioxide. For conversion of moles to grams, the number of moles is multiplied by molar mass. For example, 10 moles of water is equal to $10 \times 18 = 180g$ of water.

In the same way we can also perform conversions between grams and number of atoms or molecules and vice versa. The number of grams of potassium chloride in 100 molecules of potassium chloride is equal to $(100 \times 74.5) / 6.023 \times 10^{23} = 1.24 \times 10^{-20}g$. We can also calculate the number of atoms in a given weight of a compound. For example, how many oxygen atoms are present in 9.8 grams of sulfuric acid? We can convert grams to moles by dividing with molar mass of sulfuric acid; moles to number of molecules by multiplying with Avogadro's number and then multiplying by four since one molecule of sulfuric acid contains four oxygen atoms. So the number of oxygen atoms = $(9.8 \times 6.023 \times 10^{23} \times 4) / 98 = 2.41 \times 10^{23}$.

Density of a substance is the ratio of its mass to volume. Conversely if density is known, we can calculate the volume a certain mass will occupy or the mass a certain volume will accommodate. It is important to make sure that the correct units are used. For example, the density of 20g of sodium metal occupying a volume of 100 mL = $20/100 = 0.20$ g/cm^3 = 200 kg/m^3 in SI units. We can also calculate the mass of oxygen gas (density = 0.001g/mL) which occupies a volume of 1L at room temperature. Mass of oxygen = volume x density = 1000 mL x 0.001g/mL = 1.00g. In the same way if mass and density of oxygen gas are known, volume occupied by the gas can be calculated.

4. Cleaning and Calibration of laboratory glassware

Proper cleaning and calibration of laboratory glassware is the primary requirement for getting good results.
Cleaning glassware: A glass apparatus is considered clean, if its surface is uniformly wetted by water. The following three types of cleaning agents are usually recommended. i) Detergent: Brief soaking in warm detergent solution followed by thorough rinsing with tap water is very effective for most cleaning problems. ii) Chromic acid: Chromic acid is most effective against grease. It can be prepared by dissolving about 30g of sodium or potassium dichromate in one liter of concentrated sulfuric acid. The solution can be reused but once the solution turns green it loses its effectiveness and should be discarded. iii) A solution prepared by mixing equal volumes of 3N hydrochloric acid and alcohol is also effective, particularly for cleaning spectrophotometric cells etc. In all cases soaking for longer periods in the cleaning agent should be avoided since this may make the glass surface rough. It is important to make sure that all traces of cleaning agent are removed by thorough washing with tap water. Before use glassware must be rinsed 3-4 times with distilled or deionized water. Cleanliness of burettes and pipettes and other glassware is indicated by the absence of any water beads on the inside surface of the glassware.
Calibration of glassware: The volumetric glassware commonly employed includes a volumetric flask, burette and pipette. Each of these is provided with calibration marks and for routine laboratory work, the glassware supplied by reputed manufacturers can be assumed to deliver accurate volumes. However, in order to retain the calibration, proper care is to be exercised during their use. The following should be kept in mind. The glassware should not be subjected to high temperatures and harsh chemicals which make the glass surface rough. The standard temperature for calibration of volumetric glassware is 20^0C. Glass is a good choice for

volumetric ware since it has a low coefficient of thermal expansion. However, subjecting volumetric glassware to high temperature in an electric oven or direct flame can disturb the calibration and should be avoided.

In precise work, however, it is not safe to assume that the volume delivered is exactly equal to the amount indicated by the calibration mark. Volumetric glassware can be easily re-calibrated when required. Glassware is commonly calibrated using a liquid, the density of which is precisely known at different temperatures. Water is usually the liquid of choice. Thus a good analytical balance and distilled or deionized water at known temperature are required for calibration. Since density = mass/volume, by weighing the amount of water delivered from the glassware and dividing it by the density of water at that temperature, volume of water can be calculated. Density is affected by temperature, so it is necessary to measure the liquid temperature and look up appropriate density values. It is a good idea to keep water, balance and the glassware to be calibrated in the same room long enough to be sure everything have the same temperature. The weighing balance used should also be calibrated with standard weights from time to time.

5. Preparation of solutions

The solution preparation involves mixing solute in a smaller amount of solvent than the final volume, dissolving it completely and then diluting it to reach the final volume.

i) Calculation of the amount of solute to be dissolved: The first step is calculating the appropriate weight of solid or liquid solute to be dissolved in the solvent. The amount of Substance to be weighed (w) in grams is calculated as follows. For molar solutions, w = (Molecular weight of solute x Molarity x Total volume of solution to be prepared)/1000. For normal solutions, w = (Equivalent weight of solute x Normality x Total volume of solution to be prepared)/1000. For molecules with water of hydration molar mass of water molecules has to be included in the molar mass of the substance. For example, oxalic acid, $(COOH)_2$ $2H_2O$ has two water molecules associated with it and therefore, its molar mass is 126 and not 90. For liquid solutes of known density, since mass, m = volume x density, appropriate volume of liquid solute can be measured and dissolved in the solvent; it is not always necessary to weigh a liquid solute.

ii) Weighing method: Weighing by transfer method is more accurate, scientific and quicker than weighing an exact amount of the substance. It involves the following steps: Step 1: Weigh the empty container, Step 2: Weigh the container + substance, Step 3: Transfer the substance and weigh the container again, Step 4: Calculate molarity or normality from the weight transferred.

iii) Solution preparation: Solutions are always prepared in measuring flasks. Never use a measuring (graduated) cylinder, beaker or reagent bottle for preparing solutions of exact concentrations. Place a small funnel on the measuring flask. Transfer the substance completely very carefully. Make sure that no substance remains sticking to the weighing container and also it is not spilled out while transferring. Wash the substance into the flask using water from a wash bottle. At this stage do not fill the measuring flask completely to facilitate shaking and dissolving the substance. Make sure that all the substance has dissolved. Some substances with small aqueous solubility require thorough shaking for longer period. It is not advisable to heat the measuring flask to facilitate dissolution since heating causes error in the calibration of the flask and may also decompose the solute. When all the substance has dissolved, fill the flask up to the mark. The last few drops can be added with the help of a pipette or a dropper. Put the stopper tightly and shake to ensure complete mixing of the contents of the flask.

iv) Dilution: Diluting a more concentrated stock solution is also a commonly used method for preparing exact solutions. For dilution of stock solution or otherwise for measuring exact volumes, you can use a pipette, graduated pipette or a burette. Large volumes can sometimes be added with a measuring flask also. Do not use a measuring (graduated) cylinder or graduated beaker for this purpose. The volume of stock solution required to prepare a diluted solution can be calculated using the known relation $M_1 V_1 = M_2 V_2$ or $N_1 V_1 = N_2 V_2$; subscripts 1 and 2 stand for concentrated stock solution and diluted solution, respectively. However, these relations can only be used when concentrations are expressed on per unit volume basis; for example, molarity and normality. Using this equation with a mass-based concentration unit, such as % (w/w), leads to an error.

Some More useful Tips about solution preparation

i) It is important to learn to differentiate between rough solutions and exact solutions. Rough solutions are those solutions which have to be standardized subsequently or the solutions where exact concentration is not required. For preparing such solutions you need not waste time in weighing exact amount or measuring exact volume. Solid solutes can be weighed quickly on a rough balance while the required volume of liquid solutes can be transferred using a measuring (or graduated) cylinder. For example, there is no use taking exact weight of sodium hydroxide or measuring concentrated hydrochloric acid precisely with a pipette or micro burette for preparing sodium hydroxide and hydrochloric acid solutions, respectively. This is because sodium hydroxide and hydrochloric acid are not primary standards and therefore, even by taking exact amounts you cannot prepare exact solution. The prepared solutions have to be standardized before use.

ii) If for a certain experiment you have to weigh and transfer an exact amount, you can weigh on a watch glass or even paper provided it is butter paper or glazed paper with even edges. Transfer the whole amount using a camel hair brush directly into the measuring flask using a small funnel and then prepare the solution as usual.

iii) Do not weigh hygroscopic salts like calcium chloride and bases like sodium hydroxide on electrical or electronic balance. Doing so may corrode the metallic balance pan. Moreover, accurate weighing of such substances is meaningless and the solutions have to be subsequently standardized. They should be weighed in a glass container on a rough balance. Glazed paper cannot be used for such weighing.

iv) Substances such as sodium hydroxide, ammonium hydroxide, hydrochloric acid, acetic acid are not primary standards. Such solutions should always be standardized with a primary standard before use. For example, for standardization of bases like sodium hydroxide, oxalic acid can be used as a primary standard to know the exact concentration.

v) Ammonium hydroxide should preferably be standardized with standard HCl rather than oxalic acid because the end point in volumetric weak acid-weak base titrations is not very sharp. So in the case of ammonium hydroxide and other weak bases, the standardization involves the following steps, Oxalic acid $\rightarrow$ NaOH $\rightarrow$ HCl $\rightarrow$ NH_4OH or other weak base.

vi) It is not advisable to suck strong acids, strong bases and organic solvents by mouth. The vapours inhaled through sucking can be injurious for health and can also lead to accidents. In principle, no chemical should be sucked by mouth. A propipetter or micropipette should be employed.

vii) Use minimum amount of indicator for titrations. Larger amounts can lead to errors since indicators are also weak acids or weak bases.

viii) Use clean properly washed apparatus for good results. Rinse with distilled water before use.

6. Graph Plotting

Graph plotting enables us to visually represent data collected from an experiment. A graph is a diagram showing relation between two variables, each measured along one of a pair of axes. Graph reveals much more clearly the behavior of data such as linearity or nonlinearity, maxima and minima, points of inflexion etc. Graphs can be generated either manually or with the help of a computer. Many commercial computer programs such as Microsoft Excel, SigmaPlot etc. are available for graph plotting and in postgraduate and research labs, graph plotting is almost invariably done with the help of computer software. However, in undergraduate labs it may be the case that the students do not have access to a computer or the time to analyze the data electronically. The ability to plot a graph by hand is one of the most useful skills in physical chemistry laboratory. It is important to develop good strategies to plot graphs which are clear, tidy and accurate. Manual generation of graphs will only be discussed here.

We will briefly discuss various parts of a graph. The horizontal axis should always represent independent variable while the vertical axis represents the dependent variable. The independent variable is the parameter which is under our control while the dependent variable is the result of the experimental measurement. Points represent the collected data that are plotted on a graph. Line (either straight or curved) represents the behavior of the system. It is highly unlikely that all the points will fall directly on a line deduced from an experiment. Consequently, a line of best fit or a trend line is drawn through the points.

Tips on graph plotting*

i) Select a suitable graph paper of good quality with accurately ruled thin lines and based on data decide whether landscape or portrait graph should be plotted.

ii) The scale for each axis should be chosen in such a way the maximum possible space on the graph paper is utilized. At the same time the largest positive (and if necessary, negative) values in data should fit easily on the paper. To do this look at the maximum and minimum values of x and y; it is not necessary that they are zero at the start of the scale. Count the number of available squares and deduce a convenient scale for each square. Sufficient space should also be available for extrapolation (the extension of a line), if required. Each axis must have a linear scale, i.e. the value for each square is consistent along the axis. Labels must include an indication of what the axis represents. Usually label specifies the symbol used for the variable and its associated unit (in standard form) in brackets afterwards, for example t (s) would represent time in seconds. To avoid scale numbers that are too large or too small for convenient use, multiply the quantity by a power of 10. For example, if the values for the density, d of a gas are 0.0002, 0.0004, 0.0006, 0.0009 g/cm^3 etc., it is convenient to plot them as 2, 4, 6, 9 etc. and label the axis as d x 10^4 which is equivalent to multiplying all the scale numbers by 10^{-4}.

iii) Points should be plotted using a sharp pencil and should be marked using a symbol such as cross, circle, dot etc. If there is more than one line in the same graph the points should be given different symbols. Odd points which do not fit into the line, called Outliers (sometimes called rogue or anomalous points), should not be deleted but can be indicated by circling the point.

iv) The line to be drawn through the points will either be straight line or a curve. Straight lines should be drawn using a ruler. The line drawn should be an average line with points equidistant from the line on either side. A mathematical procedure called linear regression analysis (or least squares) can also be used to calculate the exact line. Curves drawn through the points should be smooth without any kinks. Practice the feel of the curve without touching the paper several times and then draw it smoothly, quickly and confidently. Drawing a curve in this way requires practice.

v) Graphs of scientific data should always have a title which describes the contents of the graph. The title can be written either at the top of the graph or below it. In case of difficulty in giving a title to the graph, labels used on the axes can be restated. For example, 'Plot showing change in concentration of reactant with time'.

vi) You may be required to take readings or calculate a gradient (slope) from the graph. Readings should always be taken from your line of best-fit and not from the points you have plotted. The same is true for gradient/slope calculations.

*Ref. Learning enhancement team, Student support service, University of Anglia. Website: https://portal.uea.ac.uk/student - support -service/learning - enhancement.

7. Safety in physical chemistry laboratory

i) Keep your lab working bench clean and tidy.
ii) Always wear safety glasses or other appropriate eye protection while working in the lab.
iii) Do not use mouth suction to draw up chemicals into a pipette. Use of a pipetting bulb, propipetter or suction micropipette is recommended.
iv) If some chemical is spilled on you, wash it off immediately with lots and lots of water, and then report to the teacher concerned.
v) Uncontrolled long hair or clothing (loose sleeves, ties, jewelry) that might come in contact with a flame or become entangled in mechanical equipment should be avoided.
vi) Never heat flammable materials with an open flame or near an ignition source.
vii) Do not heat or mix anything near your face (or anyone's face).
viii) Your hands are frequently contaminated while working. So do not rub your eyes with your hands.
ix) Protective clothing such as lab coat is highly recommended.
x) Do not store reagents near a sink or leave them near the balance where they can be knocked over. After use return all reagents to their proper location as soon as possible.
xi) Check all electrical equipment carefully before plugging into the power line. Do not operate electrical equipment with wet hands.
xii) No unauthorized experiments should be performed.
xiii) Check glassware before using it. Do not use broken or cracked glassware. Remember broken glass is the sharpest material known.
xiv) Never taste or smell chemicals and avoid contact with skin.
xv) Chemical waste should not be put directly in the dustbin or sink. It should be disposed of as directed by the teacher concerned.

8. Writing a Lab Report

Format of lab report: The lab report should have the following format.

i) Title of the experiment and date. The title should be brief but it should clearly indicate the content of the study.

ii) Introduction: The introduction should contain i) the objective (aim) of the experiment and ii) all the background information and theoretical basis (principle) of the experiment.

iii) Materials or Requirements: All the equipment and major chemicals required for carrying out the experiment should be listed here.

iv) Experimental Procedure: Experimental procedure should include a brief description of the experimental equipment used and procedure followed. This should be a report of what was actually done and not what was planned or what is written in the lab manual.

v) Data and observations: Data and other observations should be recorded carefully while performing the experiment. It is recommended that the numerical data obtained from the experimental procedure should be presented in tabular form wherever possible.

vi) Results: This section contains analysis of experimental data including any calculations based on the data. Graphs and charts prepared from the calculated results are also displayed here.

vii) Discussion: Discussion is the most important part of the report because that shows how well the student has understood the experiment. In the discussion section, explain, analyze and interpret your results. If your result is different from that expected, justify it by giving a suitable explanation. Calculate percentage error and mention the possible sources of error which can affect the result.

viii) Conclusions: Conclusions is mostly a single paragraph that sums up the entire experiment. Here you can compare your results with the expected or reported values.

ix) Precautions: Write only those precautions which you have actually observed during the conduct of the experiment. Do not copy them from the lab manual or text book.

Some General Tips:

i) Lab reports should be brief but comprehensive and repetition should be avoided.

ii) Reports should always be written in third person. Avoid using the words 'I' or 'we' when referring to the experimental procedure. For example, instead of 'I weighed 3.22g of the substance' the report should read '3.22g of substance was weighed'.

iii) There are some guidelines regarding the use of correct language in the report. General rules are like this: i) Experiment has already been performed and therefore use past tense when referring to the experimental procedure. For example, 'The solution was prepared...... ii) Report, equipment and theory still exist, so use present tense for them. For example, 'The objective of the experiment is.......' iii) Also report should not be written as directions such as those written in lab manual or text book. For example, do not write 'heat the solution to boiling temperature'. Instead write 'the solution was heated to boiling temperature'.

iv) It is always useful to come prepared for the lab. Before coming to the lab, read the principle of the experiment and the procedure from the lab manual or prescribed text book. This saves the precious lab time and you will be able to collect and record all the necessary information which is to be included in the final report.

v) When performing the experiment, record properly all the equipment and materials used, environmental conditions such as temperature etc. and other experimental observations. Also do not forget to include all information regarding preparation for the actual experiment such as weighing of material, solution preparation etc.

vi) Lab manual or prescribed text book should only be used as a guideline. You should not copy it word by word. Always write concisely in your own words.

vii) Raw data should be arranged properly in Tables. The number of tables should be kept minimum and each table should have a title. In the table, independent variables are written in the left hand column and the dependent variables on the right hand column. Column headings must contain all the information needed to define the values in the column. Each table and graph included in the lab report should have a number and a title. Support your results with appropriate graphs. The tips for drawing good graphs have already been discussed in the previous section.

viii) Write only what is actually observed during the experiment, not the observations you were expecting to get. Instead try to explain the observed results and suggest ways to improve the accuracy of the data. Lab reports are generally graded on the basis of your understanding of the experiment rather the final result obtained.

ix) Cite your sources: Except for general physical or numerical constants or well- known theoretical equations, any data or material taken from an outside source must be accompanied by a complete reference to that source.

x) In addition each page of report should be numbered. It is also important to mention that the hand written report should be neat and legible.

B. Section II (Viva-Voce Questions)

1. General

Q. 1. What is the difference between mass and weight?

Ans.: Though the terms weight and mass are sometimes used interchangeably, they are not the same. Mass (m) is the actual amount of material contained in a body and its SI unit is kilogram. Weight (w) is the force exerted on the object by the acceleration due to gravity. It is equal to the mass multiplied by the acceleration due to gravity (g). So weight is proportional to mass since at a given place acceleration due to gravity is constant. The SI unit of weight is newton (N). Mass is an inherent property of the matter and does not change with location whereas weight depends on the gravity at that place. Weight is measured using a scale such as a spring balance which effectively measures the pull on the mass exerted by the gravity of the earth. Mass of a body is measured by balancing it equally with another known amount of mass. It may be mentioned that wherever there is confusion in the terminology used, the guiding rule should be the units of the quantity. If the units of measurement are grams, it is mass not weight.

Q. 2. How can we covert weight to mass and vice versa? What will be the weight of a mass of 1 kg on earth surface?

Ans.: Mass and weight can be interconverted using the relation, weight (W) = mass (m) x acceleration due to gravity (g). Since the acceleration due to gravity at the surface of earth is 9.8 ms^{-2}, a mass of 1 kg on earth surface will have a weight of 9.8 N. (Newton, N is the force that gives a mass of 1 kg an acceleration of 1m s^{-2}.)

Q. 3. Define the commonly used terms gram molecular weight, molar mass and molecular mass.

Ans.: The terms, gram molecular weight and molar mass are the same and are defined as the mass of one mole of a substance having SI unit kg/mole. Molar mass is the more appropriate term and should be preferred. Molecular mass is defined as the mass of one molecule of a substance and is equal to the molar mass divided by the Avogadro's number and the SI units are kg.

Q. 4. How do you define gram equivalent weight?

Ans.: Gram equivalent weight is the mass in grams of one equivalent and is defined classically as the mass of an element/compound/ion which combines with or displaces 1 part of hydrogen or 8 parts of oxygen or 35.5 parts of chlorine by mass. However, it is not always possible to apply this classic definition to determine the equivalent weights of different chemical species. Following are more workable definitions of equivalent weight.

Gram equivalent weight of an element = Gram atomic weight / valency. For example, atomic weights of sodium and calcium are 23 and 40, respectively. The corresponding equivalent weights are 23/1 = 23 and 40/2 = 20, respectively.

In acid-base reactions, equivalent weight is the mass of a given substance that will supply or react with one mole of hydrogen ions. Accordingly, Gram equivalent weight of an acid = Gram molecular weight /Basicity of acid. The basicity of an acid is equal to ionizable hydrogen ion (proton) in its aqueous solution. For example, gram equivalent weights of HCl with basicity 1 is 36.5/1 = 36.5 while that of H_2SO_4 with basicity 2 is 98/2 = 49.

Gram equivalent weight of a base = Gram molecular weight / Acidity of base. The acidity of a base is equal to ionizable hydroxyl ion (OH^-) in its aqueous solution. For example, acidity of potassium hydroxide (KOH, Molecular weight = 56) is 1 while that of calcium hydroxide ($Ca(OH)_2$, Molecular weight = 74) is 2. Therefore, equivalent weight of KOH = 56/1 = 56 and that of $Ca(OH)_2$ = 74/2 = 37.

Gram equivalent weight of an ion is equal to the molecular weight of ion divided by the number of electronic charges on the ion. For example, the number of electronic charges on carbonate ion (CO_3^{2-}, molecular weight = 60) are 2 and the equivalent weight of CO_3^{2-} ion = 60/2 = 30.

In a redox reaction, one of the reacting species is oxidizing agent and the other is reducing agent. Gram equivalent weight is the mass of a given substance that will supply or react with one mole of electrons in a redox reaction. Thus gram equivalent weight of an oxidizing or reducing agent = Gram molecular weight / Number of electrons transferred in the redox reaction.

Q. 5. What is meniscus (liquid) and why it is important for measuring volumes?

Ans.: A meniscus is a curved surface at the top of a column of liquid caused by the surface tension of the liquid. It can be concave or convex depending on the liquid and the surface of the container. A concave meniscus occurs when the molecules of the liquid are more strongly attracted to the container than to each other. Most liquids, including water, present a concave meniscus. A convex meniscus is produced when the molecules of the liquid are more strongly attracted to each other than to the container. A familiar example of a convex meniscus is mercury in a glass container. The bottom of the meniscus is called the lower meniscus while the top of the meniscus is the upper meniscus. For concave meniscus we should always read the lower meniscus while for convex meniscus upper meniscus is read. In liquids with concave meniscus, only when the solution is so dark or opaque that the lower meniscus in not clearly visible, upper meniscus is read.

Q. 6. What is parallax error and how parallax error can be avoided?

Ans.: Parallax is a displacement or difference in the apparent position of an object viewed along two different lines of sight. For example, while taking a burette reading, the apparent position of the liquid meniscus appears altered as the line of vision is raised or lowered. This is called Parallax error and occurs if the eye is not at the same level as the liquid meniscus. To avoid parallax error, the position of the eye should be at the same level as the liquid meniscus. These precautions should also be observed while reading other volumetric glassware of similar character.

Q. 7. What is the difference between a laboratory thermometer and a clinical thermometer?

Ans.: There are the following two basic differences between a laboratory thermometer and a clinical thermometer. A laboratory thermometer can measure temperature over a wide range while a clinical thermometer can only measure temperature in the range 35^0C to 42^0C and thus cannot be used to measure temperatures higher and lower than this. Secondly clinical thermometer has a U bend tube called kink which stops mercury's back flow into the bulb after it is removed from patient's mouth while there is no bend or kink in laboratory thermometer.

Q. 8. Like mercury thermometer, why not a water thermometer?

Ans.: Water is not a suitable liquid for use in a thermometer due to the following reasons. i) The coefficient of thermal expansion of mercury is much larger than that of water. Moreover, it varies linearly with temperature whereas that of water varies non-linearly. ii) Very low freezing point ($- 37^0C$) and very high boiling point (356^0C) makes it a suitable fluid to measure both positive and negative temperatures. With water it is not possible to measure temperatures below $0°$ and above $100°C$. iii) Easier to take reading using mercury since it has a silvery colour while water is transparent. iv) The condensation of water vapours on the walls of capillary tube can also be a source of error.

Q. 9. What are the relative advantages and disadvantages of alcohol thermometers as compared to mercury thermometers?

Ans.: Relative advantages of alcohol thermometers as compared to mercury thermometers are low cost, less hazardous in case of breakage, large coefficient of expansion of alcohol and low freezing point (- $115°C$). Alcohol thermometers are thus useful for measuring very low temperatures. However, alcohol thermometers cannot be used to measure high temperatures due to low boiling point of alcohol. Moreover, alcohol is colourless, wets the walls of the tube and can polymerize. Alcohol thermometers are also less durable than mercury thermometers. Large coefficient of expansion is also a disadvantage if small length compact thermometers are required.

Q. 10. Why numerical values of density and specific gravity are nearly same although they are defined differently?

Ans.: Density is mass of unit volume of liquid and is an absolute quantity whereas specific gravity of a liquid is a relative quantity, defined as the ratio of the density of liquid to that of a reference material, usually water. However, because the density of pure water is so close to 1, specific gravity and density have nearly the same numerical value so long as the density is given in g/cm^3. Density is very slightly less than specific gravity.

Q. 11. What is the effect of temperature on density and specific gravity of liquids?

Ans.: Both density and specific gravity change with temperature. Since density is mass per unit volume and volume increases with increase in temperature, density decreases with increase in temperature. In the case of specific gravity, if temperature changes the density would change for both the substance and water. But the reason why the specific gravity does not stay the same is that the degree of change in volume with change in temperature is different for different substances?

Q. 12. Why special care is needed in handling volatile and hygroscopic substances?

Ans.: A substance with higher vapour pressure vaporizes more readily than a substance with a lower vapour pressure. Highly volatile substances cannot be left open or uncovered in the atmosphere. In order to avoid errors in measurements, they should always be kept properly covered and special care is needed to avoid their evaporation during use. Also one has to be very careful that the vapours are not inhaled accidently.

A hygroscopic substance readily attracts water from the atmosphere by either adsorption or absorption. Salts such as calcium chloride, sodium hydroxide are so hygroscopic that they readily dissolve in the absorbed water. This property is called deliquescence. Therefore hygroscopic salts should also be properly covered during handling. Such substances cannot be weighed accurately and the prepared solution has to be standardized before use.

Q. 13. What is the difference between aliquot and diluent?

Ans.: An aliquot is the sub-volume of original sample whereas a diluent is the material with which the sample is diluted.

Q. 14. What is a buffer solution?

Ans.: A buffer solution is one which resists the change in pH when a small amount of acid or base is added to it. An aqueous buffer solution usually consists of the mixture of a weak acid and its salt or a weak base and its salt.

Q. 15. What is valency?

Ans.: Valency is a measure of the combining capacity of an atom to form chemical bonds. Valence is defined by the IUPAC as the maximum number of univalent atoms (hydrogen or chlorine) that may combine with an atom of the element under consideration. Atoms are known to be capable of displaying more than one valence.

Q. 16. What is the difference between activity and concentration what is the relation between the two?

Ans.: Concentration signifies the actual amount of substance used to prepare solution while activity is the effective concentration of solution. Due to non-ideality of solutions, the total amount of the added substance is not available for reaction. Activity is the amount which is actually available. Since ideality can only be approached, no solution is truly ideal. Hence activity should always be used.

When dealing with very dilute solutions it is sometimes assumed that activity is approximately equal to concentration because an infinitely dilute solution tends to behave ideally. Activity (a) and concentration (c) are related to each other by the expression, $a = \gamma c$, where γ is the activity coefficient. For an ideal solution, activity coefficient, γ is unity and activity = concentration. In dilute solutions, as the concentration of solution, $c \rightarrow 0$, activity coefficient, $\gamma \rightarrow 1$ and activity, $a \rightarrow$ concentration, c. Since for very dilute solutions $\gamma \approx 1$, $a \approx c$, activity $\approx$ concentration.

Q. 17. What do you understand by STP?

Ans.: The abbreviation STP in chemistry stands for standard temperature and pressure and is generally used in calculations involving gases. According to the recommendations of the International Union of Pure and Applied chemistry (IUPAC), standard temperature is equal to 0°C (= 273.15 K) and standard Pressure is 1 atm (= 101.3kPa or 760 mm Hg). At STP 1 mole of any gas occupies a volume of 22.4L.

Q. 18. What is standard state? Is there any relation between **standard temperature and pressure (STP) and standard state?**

Ans.: Standard temperature and pressure (STP) should not be confused with the standard state. Standard state is the reference state for the material's thermodynamic properties. The standard state for a gas is the hypothetical state where the gas obeys ideal gas law at a pressure of 10^5 Pa ($\approx$ 1 atmosphere). The standard state for liquids and solids is simply the state of the pure substance subjected to a total pressure of 10^5 Pa. Strictly speaking temperature is not part of the definition of standard state. However, thermodynamic quantities are usually tabulated at a specific temperature most commonly 298.15K (25.00°C).

Q. 19. What is the difference between STP and NTP?

Ans.: STP stands for standard temperature and pressure and has already been defined in Q. 17. NTP stands for normal temperature and pressure. The most used NTP values are those defined by National Institute of Standards and Technology (NIST) according to which normal temperature is 20°C (= 293.15 K) and normal pressure is 1 atm (= 101.3kPa or 760 mm Hg). At NTP 1 mole of a gas occupies a volume of 24.0 L. NTP conditions allow comparisons to be made between data obtained from different laboratories.

Q. 20. What is the meaning of limiting and excess reagent in chemical reactions?

Ans.: In a chemical reaction, the substance which is totally consumed when the reaction is complete is called the limiting reagent (or limiting reactant). Chemical reactions proceed according to the mole ratio of the reactants in the balanced chemical equation. Limiting reactant determines the amount of product formed since reaction cannot proceed further without it. On the other hand, the reactant which still remains or the reactant which is not completely used up in reaction is called excess reagent. Limiting reagent can be identified from the known stoichiometry of the chemical reaction.

Q. 21. What do you understand by scientific notation for writing numerical values?

Ans.: Scientific notation is useful in writing very small and very large numbers. Scientific notation is the exponential notation in which any number can be represented in the form $N \times 10^n$ where exponent n is an integer having positive or negative values and N is a number (called digit term) which varies between 1.000 and 9.999. For example, we can write number 4532 as 4.532×10^3 and 0.0006391 as 6.391×10^{-4}. For very small and very large numbers it is always desirable to use scientific notation. For performing mathematical operations such as multiplication and division on numbers expressed in scientific notation, follow the same rules which are there for exponential numbers. For addition and subtraction, it is preferable to first write the numbers in such a way that they have the same exponent and then add or subtract.

Q. 22. What is the difference between interpolation and extrapolation of data?

Ans.: Interpolation is the prediction of data points that fall within the range of the measured data that is between the existing data points. Extrapolation, on the other hand is the prediction of data points from beyond the range of the measured data.

Q. 23. Why the exact amount of a thick (highly viscous) liquid should preferably be taken by weight rather than by volume?

Ans.: When thick (highly viscous) liquids are measured with a pipette, lot of liquid is left adhering to the sides of the pipette and consequently the volume delivered is less. It is therefore, preferable to take it by weight. The liquid can be weighed by putting it in a pre-weighed container. After transferring the liquid to the measuring flask, the container should be rinsed several times with solvent to ensure complete transfer.

Q. 24. What is the meaning of lag time of the instrument?

Ans.: Most instruments require some time, called lag time, to reach equilibrium after they are switched on. Taking a reading before the instrument is stable will result in faulty measurement. Usually a time gap of 15-20 minutes is required for most of the instruments to warm up and stabilize. Therefore, readings should be taken only after the instrument has stabilized.

Q. 25. How can you categorize a salt as sparingly soluble?

Ans.: In general, salts are divided into the following categories according to their solubility.

i) Soluble if solubility is greater than 0.1 M. ii) Slightly soluble if solubility lies between 0.01 and 0.1 M. iii) Sparingly soluble if solubility is less than 0.01 M.

Q. 26. What is solubility product of a salt?

Ans.: The solubility product constant, K_{sp} is the equilibrium constant for the dissolution reaction of a salt in water. Solubility product concept is usually used for salts which are sparingly soluble in a solvent. In the saturated solution, the concentration of solid salt present is taken as unity. It is defined as the mathematical product of its dissolved ion concentrations raised to the power of their stoichiometric coefficients. For example, solubility product of silver chloride, $AgCl = [Ag^+][Cl^-]$ while that of silver chromate, Ag_2CrO_4 is $[Ag^+]^2 [CrO_4^{2-}]$.

Q. 27. What is the difference between solubility product and ionic product of a salt?

Ans.: The basic difference is that the solubility product is always defined for saturated solution whereas ionic product is applicable to all type of solutions of any concentration. For a given salt solubility product is constant at a given temperature while ionic product changes with change in the concentration of solution.

Q. 28. What is the significance of the solubility product constant of a salt?

Ans.: i) The smaller the solubility product of a substance, the lower is its solubility. ii) Knowledge of solubility product enables determination of solubility of a sparingly soluble salt. iii) Solubility product is useful in predicting whether on mixing two ionic solutions, precipitation will occur or not. If ionic product exceeds solubility product, precipitation will occur in solution. iv) The concept of solubility product is also useful for separation of basic radicals into different groups in qualitative analysis and in the purification of some salts.

Q. 29. What is common ion effect?

Ans.: Weak acids and weak bases are not fully ionized and in solution and equilibrium is established between the ions formed and the unionized acid/base. The phenomenon of suppression of the degree of dissociation of a weak acid or a weak base by the addition of a strong electrolyte containing a common ion is known as common ion effect.

Q. 30. What is critical solution (consolute) temperature? Define upper consolute temperature and lower consolute temperature.

Ans.: Critical solution temperature also called consolute temperature is the temperature at which two partially miscible liquids attain complete miscibility as the temperature is raised or lowered. The upper cosolute temperature is the critical temperature above which the components of a mixture are miscible in all proportions. Similarly lower consolute temperature is the critical temperature below which the components of a mixture are miscible in all proportions. For example, at a concentration of 11% by weight of phenol a partially miscible phenol-water mixture at room temperature becomes completely miscible and exists as a single phase at and above 65.8^0 which is the upper consolute temperature of the system. It is difficult to determine the lower consolute temperature as one of the components may freeze at that temperature. In general, miscibility and hence the consolute temperature also depend on the presence of a third component such as impurity.

2. Accuracy, Precision and Error Analysis

Q. 1. What is an experimental error?

Ans.: No physical quantity can be measured with perfect certainty; there are always errors in any measurement. Experimental error is the difference between the experimental value and the actual value of a property.

Q. 2. What do you understand by uncertainty and how it is expressed?

Ans.: Uncertainty defines how far an experimental measurement is from the best estimate or the true value. So measurement = best estimate ± uncertainty. For example, a measurement of 4.78 ± 0.03g means that the actual mass of the substance lies between 4.75g and 4.81g.

Q. 3. What is the difference between error and uncertainty?

Ans.: Error is the difference between the measured value and the true value of a parameter whereas uncertainty characterizes a range of values within which the true value lies.

Q. 4. What are the different ways by which uncertainty in an experimental measurement can be expressed?

Ans.: Uncertainty in measurement can be expressed either as absolute uncertainty or relative uncertainty which is the ratio of absolute uncertainty and the measured value. Multiplying the relative uncertainty by the measured value yields the absolute uncertainty. Percentage uncertainty is equal to relative uncertainty multiplied by hundred. For example, if the best estimate of length is 2.24 cm but due to uncertainty the measured length is 2.22 cm or 2.26 cm, the absolute uncertainty = 0.02 cm and length = 2.24 ±

0.02 cm, relative uncertainty = (0.02/2.24) = 0.0089 and percentage uncertainty = (0.02/2.24) x 100 = 0.89% and length = 2.24 ± 0.89%. Absolute uncertainty has the same units as the measured value with which it is associated while relative uncertainty is always unit less. The final result is usually expressed in terms of absolute uncertainty or percentage uncertainty.

Q. 5. What are the basic rules of uncertainty calculations?

Ans.: Rule 1: During both addition and subtraction, absolute uncertainties are added. If ΔA and ΔB are absolute uncertainties in A and B, respectively then

$(A \pm \Delta A) + (B \pm \Delta B) = (A + B) \pm (\Delta A + \Delta B)$ and
$(A \pm \Delta A) - (B \pm \Delta B) = (A - B) \pm (\Delta A + \Delta B)$

For example, $(5.40 \pm 0.02) + (3.80 \pm 0.03) = (9.20 \pm 0.05)$ and $(5.40 \pm 0.02) - (3.80 \pm 0.03) = (1.60 \pm 0.05)$. Please note that even when subtracting the measured values, the absolute uncertainties for the two measurements have to be added in the final result.

Rule 2: During multiplication and division, the absolute uncertainties for each measurement have to be first converted to percentage uncertainties which are then added to get the percentage uncertainty in the result. The total percentage uncertainty in result can again be converted to absolute uncertainty, if required. If ΔA and ΔB are the absolute uncertainties in A and B respectively, convert them to percentage uncertainties ε_A and ε_B and then

$(A \pm \varepsilon_A) \times (B \pm \varepsilon_B) = (A \times B) \pm (\varepsilon_A + \varepsilon_B)$ and
$(A \pm \varepsilon_A) \div (B \pm \varepsilon_B) = (A \div B) \pm (\varepsilon_A + \varepsilon_B)$

We illustrate this with the following example. In the multiplication, $(4.40 \pm 0.03) \times (2.00 \pm 0.01)$, 0.03 and 0.01 are absolute uncertainties. In order to calculate the uncertainty in the final result we have to first convert them to percentage uncertainties as follows.

Percentage uncertainty in A = (0.03/4.40) x 100 = 0.68. Similarly percentage uncertainty in B = (0.01/2.00) x 100 = 0.50. The total percentage uncertainty = 0.68 + 0.50 = 1.18. Therefore, $(4.40 \pm 0.03) \times (2.00 \pm 0.01) = (8.80 \pm 1.18 \%)$. Now we can convert the percentage uncertainty to absolute uncertainty as follows: $(1.18/100) \times 8.8 = 0.10$. So

$(4.40 \pm 0.03) \times (2.00 \pm 0.01) = (8.80 \pm 0.10)$. The same procedure applies to division.

Rule 3: For a number raised to a power, multiply the percentage uncertainty by the power; $(A \pm \varepsilon_A)^n = (A^n \pm n\varepsilon_A)$. The percentage uncertainty value can then be converted to absolute uncertainty, if required.
For example, Cube: $(2.00 \pm 1.0\%)^3 = (8.00 \pm 3.0\%) = (8.00 \pm 0.24)$
Square root: $(25.00 \pm 1.0\%)^{1/2} = (5.00 \pm 0.5\%) = (5.00 \pm 0.02)$

Rule 4: For multiplying a number by a constant, the rules are different when dealing with absolute uncertainties and percentage uncertainties. If we multiply the measured value with absolute uncertainty by a constant, the absolute uncertainty is also multiplied by the constant. If c is the constant, $c(A \pm \Delta A) = cA \pm c(\Delta A)$. For example, $4.0(2.00 \pm 0.01) = (8.00 \pm 0.04)$. However, the percentage uncertainty does not change. It is not multiplied by the constant; $c(A \pm \varepsilon_A) = cA \pm \varepsilon_A$. For example, $3.4(2.0 \pm 1.0\%) = (6.8 \pm 1.0\%)$.

Q. 6. What is the difference between precision and accuracy?

Ans.: In experimental science precision and accuracy convey different meanings. Accuracy refers to the extent to which a given measurement is close to the actual or known value whereas precision refers to the extent to which the repeated measurements of the same quantity are close to each other. For a measurement to be free from errors, it should have high level of accuracy as well as precision. Accuracy is related to systematic errors while precision is related to random errors.

Q. 7. Which parameter determines the accuracy of any measuring instrument?

Ans.: The accuracy of any measuring instrument is limited by its least count. Least count error results due to limited resolution of the instrument. The smaller the least count of an instrument, the more accurately it can measure a physical quantity. No instrument can accurately measure a value smaller than its least count. A ruler, for example, has centimeter and millimeter markings and therefore, the smallest length which can be accurately measured is 1 mm, i.e. the least count of the ruler is 1mm. If the length measured with a ruler is stated as 7.8 mm, 0.8 is just a guess work. However, it can be measured accurately using a vernier caliper with least count 0.1 mm.

Q. 8. How instrumental error is related to least count?

Ans.: The instrumental error actually incurred may not be as large as the least count but it is not possible to assess exactly how much it is. For example, if the least count is 1 mm and the measured length is between 7 and 8 mm, it is unacceptable to report it as 7 mm or 8 mm since we are sure that the actual value is more than 7 but less than 8 mm. However, it is easy to judge whether it is closer to 7mm or 8mm. The more precise and accurate measurement would be the estimated value ± some agreed level of anticipated error. It is unlikely that the measurement could be incorrect by more than half the least count. So the generally accepted level of error in reading the smallest division is considered half the least count. If for example, the estimated value is 7.3, it should be reported as 7.3 ± 0.5 mm. The smaller the least count of the instrument therefore, the smaller is the instrumental error incurred.

Q. 9. What about the sign of the experimental error?

Ans.: In any measurement, Error = Experimentally measured value − True or known value. Thus the error would be positive if the measured value is more than the true value and negative if the measured value is less than the true value.

Q. 10. What is the difference between absolute error, relative error and percentage error?

Ans.: Absolute error is defined as the absolute value (or magnitude) of the difference between measured and true or known value. The relative error is the absolute error divided by the true or known value. The percentage error is the relative error multiplied by 100. For example, if the measured length is 38.00 cm and the true length is 37.50 cm, the absolute error is 0.50 cm, the relative error is $0.50/37.50 = 0.01$ and the percentage error is 1%.

Q. 11. Five measurements for the volume of an object were 44.0, 43.8, 43.7, 43.3 and 43.2 m³. If the actual or true volume of the object is 43.4 m³, calculate the average experimentally measured volume, absolute error, relative error and the percentage error.

Ans.: Average experimentally measured volume = (44.0 + 43.8 + 43.7 + 43.3 + 43.2)/5 = 43.6 m³. Absolute error = (43.6 − 43.4) = 0.2 m³. Relative error = 0.2/43.4 = 0.005. Please note that the relative error should be reported only up to one significant figure since the value 0.2 used in the calculation has only one significant figure. Percentage error = 0.5%. (The concept of significant numbers and the rules for rounding off numbers are discussed in detail subsequently in Q. 14 – 17).

Q. 12. The actual length of an object is 20 cm while the measured length has 10% error. What is the experimentally measured length of the object?

Ans.: Percentage error = {(Experimentally measured value − Actual value)/Actual value} x 100 = {(Experimentally measured value -20)/20} x 100 = 10. Experimentally measured length = 22 cm.

Q. 13. What is the correct way to report an experimental measurement?

Ans.: It is incorrect to report that part of the experimental value which is a guess work. For example, if the least count of the instrument is 1mm, the measured length should be reported as say 15 mm. It should be reported as 15.0 mm only if the least count is 0.1 mm and 15.00 if the least count is 0.01 mm. Since the instrumental error is usually equal to half the least count, a value stated as 15 mm means that it lies anywhere between 14.5 and 15.5 mm. Similarly if the least count is 0.1 mm, a value stated as 15.0 means that it lies anywhere between 14.95 and 15.05 and if the least count is 0.01 mm, a value stated as 15.00 means that it lies anywhere between 14.995 and 15.005.

Q. 14. In scientific data numbers 15, 15.0 and 15.00 are not same. Why?

Ans.: Although all the three numbers have the same value, they convey different meanings about how precise they are. The numbers 15, 15.0 and 15.00 have two, three and four significant figures, respectively. The number 15.0 is 10 times more precise than 15 and number 15.00 is 100 times more precise than 15.

Q. 15. What do you understand by significant figures?

Ans.: Significant figures are the digits in a number that express the precision of an experimentally measured or calculated value. While reporting experimental data, only those digits which are consistent with the resolution of the measuring devise should be retained in the final result. The precision of the final result cannot be more than that of the instrument used for experimental measurement. For example, false precision is created if length measured with a ruler is reported to be 8.4379 cm. It should be written as 8.4 cm since the least count of the ruler is 0.1cm. In science, only those numbers that have significance (derived from measurement) are written. Similarly in scientific calculations the value with the least number of significant figures decides the number of significant figures to be retained in the final result.

Q. 16. Are there any rules for determining how many significant figures are in a number? Explain with examples.

Ans.: Following rules are helpful for determining how many significant figures are in a number.

i) Non-zero digits are always significant. For example, 67 has two significant figures; 6 and 7 while 23781 has five significant figures; 2, 3, 7, 8 and 1.

ii) Any zeros between two significant digits are significant. For example, 203.5807 has seven significant figures; 2, 0, 3, 5, 8, 0 and 7.

iii) Leading zeros are never significant. For example, 0.0053 has two significant figures; 5 and 3. Upon writing the number in scientific notation, 5.3 x 10^{-3} the non-significant zeros will disappear.

iv) Trailing zeros are significant only if the number contains a decimal point. That is the zeros to the right side of non-zero digit in the decimal portion only are significant. For example, number 81.00 has four significant figures and 0.00430 has three significant figures; 4, 3 and 0. Trailing zeros indicate up to which decimal point the measurement is precise.

v) In a number without a decimal point, trailing zeros may or may not be significant. For example, just by looking at the number such as 4500, it is not clear whether it is precise or it has been rounded to the nearest hundred.

Q. 17. What is the correct way of rounding an experimentally determined value?

Ans.: The basic concept of significant numbers is often used in connection with rounding. Rounding is usually done to a certain number of significant figures based on the precision of the experimental data. The result cannot be more precise than the least precise value in the data. For example, a chain cannot be stronger than its weakest link. In a team race, the speed of the team is dictated by the slowest member of the team. In the same way, the answer cannot be more precise than the least precise measurement. The following rules provide guideline for rounding of numbers.

Rule 1: If the digit to be dropped is greater than 5, then add 1 to the last digit to be retained and drop all digits farther to the right. For example, 4.368 should to be rounded to 4.37 if we need three significant figures in measurement and 4.4 if we need two significant figures in measurement.

Rule 2: If the digit to be dropped is less than 5, then simply drop it without adding any number to the last digit. For example, 5.332 should be rounded off to 5.33 if we need three significant figures in measurement and 5.3 if we need two significant figures in measurement.

Rule 3: If the digit to be dropped is exactly 5 then: (A) If the digit to be retained is even, then just drop the number 5. For example: 7.465 is rounded off to 7.46 if we need three significant figures in measurement and

2.45 is rounded off to 2.4 if we need two significant figures in measurement. (B) If the digit to be retained is odd, then add "1" to it. For example: 9.335 is rounded off to 9.34 if we need three significant figures in measurement and 8.35 is rounded off to 8.4 if we need two significant figures in measurement.

Remember: zero is considered an even number. 4.05 should be rounded off to 4.0, if we need two significant figures in measurement.

Example: If number 26.36205 is rounded to different number of significant figures, the following rounded numbers are obtained.

2 significant figures – 2 and 6 are first two significant figures and the next number is 3 which is less than 5 and therefore, the number is rounded off to 26.

In the same way, rounded to 3 significant figures, it is 26.4; rounded to 4 significant figures, it is 26.36; rounded to 5 significant figures, it is 26.362; rounded to 6 significant figures, it is 26.3620.

Q. 18. In mathematical operations involving significant figures, what is the correct way to report the answer?

Ans.: In mathematical operations involving significant figures, the answer should be no more precise that the least precise number used to get the answer. The rules described below are different for addition/subtraction and multiplication/division.

i) Addition and Subtraction: When doing addition and subtraction with measured values we have to look at the decimal portion (right side of decimal point) of the numbers only. Add and subtract the numbers in the usual manner and round the answer to the minimum number of decimal places in the decimal portion of any number in the problem. For example, the calculated answer for 48.28 + 0.5 is 48.78 but the correct answer is 48.8.

ii) Multiplication and Division: When doing multiplication and division with measured values, the result is to be rounded to the same number of significant figures as the value with the least number of significant figures in any number in the problem. For example, the calculated answer for 5.350 x 3.12 is 16.692 but the correct answer is 16.7 with three significant figures as one of the numbers used in the calculation (3.12) has only three significant figures.

Q. 19. Identify the number of significant figures in the following numbers.
   6.0500, 0.00357, 5.03 x 10⁻⁵, 0.004008

Ans.: Number 6.0500 has five, 0.00357 has three, 5.03 x 10^{-5} has three and 0.004008 has four significant figures in accordance with the rules for counting the number of significant figures. For a number written in scientific notation such as 5.03 x 10^{-5}, only significant figures in the numerical portion are counted.

Q. 20. If an object has a mass of 17.532 kg and a volume of 12.2 m³, calculate the density of the object.

Ans.: Density = Mass/Volume = 17.532/12.2 = 1.43704918 kg m^{-3}. This is the calculated value which cannot be reported like this. The answer can have no more than three significant figures since the least accurate number (12.2) in the calculation contains only three significant figures. Therefore, the correct answer is 1.44 kg m^{-3} after rounding off. Please remember that rounding should be done only at the end of the calculation.

Q. 21. A drug sample contains 5.73 g of compound A, 2.5008 g of compound B and 6.6 g of compound C. What is the total weight of the drug sample?

Ans.: Total weight of drug sample = 5.73 + 2.5008 + 6.6 = 14.8308 g. However, the result should be reported to the same number of decimal places as the number with the fewest decimal places. Since one of the numbers is known only up to one decimal place, the correct answer is 14.8 g.

Q. 22. What is the volume of an object of length 23.673 m and area 5.12 m²?

Ans.: Volume of object = Length x Area = 23.673 x 5.12 = 121.20576 m³. Since area 5.12 m² has only three significant figures, the correct answer is 121 m³.

Q. 23. How many types of experimental errors can be involved in a measurement?

Ans.: The experimental errors can be classified into three broad categories: systematic errors, random errors and blunders.

Q. 24. What are systematic errors? Explain with examples.

Ans.: Systematic errors are reproducible errors which occur consistently in the same direction. Such errors persist throughout the experiment and produce results that are consistently high or low. Systematic errors are usually difficult to identify and correct but can be minimized by careful analysis of the design of the experiment and the equipment used. Systematic errors can be of the following three types.
i) Instrumental errors: Instrumental errors occur due to some problems associated with the measuring instruments. For example, a poorly calibrated instrument such as a thermometer that reads 102°C when immersed in boiling water and 2°C when immersed in ice-water bath at atmospheric pressure will give consistently high results. Such errors are also called zero errors.
ii) Environmental errors: Environmental errors occur due to some external conditions in the laboratory such as variation in temperature, pressure or humidity during the course of the experiment. For example, temperature fluctuations may influence various physico-chemical properties such as conductance, viscosity etc.
iii) Theoretical errors: Theoretical errors occur due to the approximations used in deriving the equations for the model system employed.

Q. 25. What are random errors? Explain with examples.

Ans.: Random errors, also called precision errors are random fluctuations in the measurements caused by unidentifiable causes. Random errors affect the precision of a measurement. These occur primarily due to the lack of precision of the measuring devise and random fluctuations in experimental conditions. There might be some other parameters also which affect measurement but are beyond our control. Such errors can also occur due to wrong observation or reading taken. For example, there can be a parallax error in reading a meter scale. The perturbations occur on either side of true value and therefore, such errors can be reduced by taking the average of a large number of readings and can be evaluated by statistical analysis of data.

Ref. Book: Measurement and Instrumentation. Theory and Application. By Alan S. Morris and Reza Langari. Academic Press. Elsevier, New York, 2012, p. 39.

Q. 26. Give a brief account of the statistical analysis of random errors.

Ans.: Random errors can be evaluated by statistical analysis of data. If a given measurement, x is repeated N times, the best estimate of the actual value is the average or mean of all the measurements, $\bar{x}$.

$$\bar{x} = \frac{\sum_{i=1}^{N} x_i}{N}$$

Although average or mean is the best available estimate, it is not exact. To approach a true mean value, averaging has to be done for an infinite number of measurements. For a relatively small number of measurements, the average deviation, $\Delta \bar{x}$ can be calculated. Average deviation gives information concerning how much the measured values deviate from the mean. To do this, the deviation of a particular measurement from the mean (also called a residual), $x_i - \bar{x}$ is calculated.

$$\Delta \bar{x} = \sqrt{\frac{1}{N} \sum_{i=1}^{N} |x_i - \bar{x}|}$$

The absolute values of the deviations from mean are summed up and divided by the total number of measurements to get the average deviation, $\Delta \bar{x}$. Absolute values of deviation are considered because the deviations are both positive and negative since they are on either side of the mean and if sign of deviation is considered, the sum will always be zero. However, the average deviation $\overline{\Delta x}$, calculated using absolute values of deviation does not give any information about the direction of error incurred. Another parameter is the relative average deviation, which is the average deviation divided by the mean ($\overline{x}$). The relative average deviation can also be expressed as percentage by multiplying the relative average deviation with 100.

Mean square deviation and standard deviation are much better ways of expressing the precision of the measurements. To calculate mean square deviation (MSD) also called variance (V), each of the N deviations of individual measurements from the mean are squared and the sum of these is divided by N.

$$MSD \text{ or } V = \frac{1}{N}\sum_{i=1}^{N}(x_i - \overline{x})^2$$

The square root of this quantity is the standard deviation, σ_x.

$$\sigma_x = \sqrt{\frac{1}{N}\sum_{i=1}^{N}(x_i - \overline{x})^2}$$

Any single value has an uncertainty equal to the standard deviation also sometimes known as the root-mean-squared (RMSD) deviation. Such statistical analysis applies only to the errors that follow the normal (or Gaussian) distribution. A normal distribution of data means that most of the readings in a set of data are close to the average, while relatively few readings tend to one extreme or the other. Normal distribution occurs naturally in many situations and is characterized by a bell-shaped curve which is symmetric at the center i.e., around the mean value, $\overline{x}$. Standard deviation σ_x represents the uncertainty in a single measurement. However, the average of a set of data values has much smaller uncertainty. We also know that mean ($\overline{x}$) gives true value of measured property only if averaging is done over an infinite number of measurements. For a finite set of measurements, another useful statistical parameter is the standard deviation of the mean or standard error of the mean, $\sigma_{\overline{x}}$ defined as

$$\sigma_{\overline{x}} = \frac{\sigma_x}{\sqrt{N}}$$

Since N represents the number of measurements, $\sigma_{\overline{x}}$ tends towards zero as the number of measurements in the data set tends to infinity. The standard

deviation of the mean is related to the uncertainty in the mean or average value. It is easy to see that more the number of data values, the smaller the uncertainty will be. For a series of measurements of a given quantity x with independent and random errors, the best estimate of the value can be expressed as
$$x = \bar{x} \pm \sigma_{\bar{x}}.$$

Q. 27. What is Bessel correction factor and how it modifies the expressions for statistical evaluation of data?

Ans.: The mean square deviation and the standard deviation defined in previous question are based on the assumption of an infinite set of data values. Since we cannot have infinite data set, the calculated value of mean is different from true mean. In other words, there is an error in the average or mean ($\bar{x}$) used in the calculation of mean square deviation and standard deviation. Bessel correction factor tends to reduce the bias due to finite sample size. The unbiased estimate of variance and standard deviation is obtained by multiplying the uncorrected values by Bessel correction factor (= N/(N - 1)). The modified expressions are given below.

$$MSD \ or \ Variance = \frac{1}{N-1} \sum_{i=1}^{N-1} (x_i - \bar{x})^2$$

$$Standard \ deviation, \ \sigma_x = \frac{1}{N-1} \sqrt{\sum_{i=1}^{N-1} (x_i - \bar{x})^2}$$

Q. 28. What are blunders?

Ans.: Another source of error is called a blunder. Blunder is actually a mistake committed by the person carrying out the experiment. For example recording a wrong value, misreading a scale, forgetting a digit while reading a scale, faulty recording of measurements etc. are all classified as blunders. These blunders can usually be taken care of by careful repetition of the experiment or by requesting another student to check your measured and recorded experimental data. Blunders can be easily eliminated and therefore, should not be included in the analysis of data.

Q. 29. What do you understand by instrument resolution and null difference method of measuring a quantity?

Ans.: The precision of a measurement also depends on the measurement resolution of the instrument used. Measurement resolution is defined as the smallest change in the underlying physical quantity to which the instrument responds. For example, the resolution of a weighing scale is the smallest increment in the added weight that can be detected or displayed on the screen. Same is true for various other digital instruments. It is important to have knowledge of the resolution of the measuring devise.

Null type instruments in which zero or null indication determines the magnitude of the measured quantity give more precise measurements. They use null difference method. The measured signal is the difference between a known quantity and the quantity being measured. The known quantity is varied until the null point is attained, that is, the difference between the two is reduced to zero. The most common example in physical chemistry lab is a potentiometer which works on the poggendorf's compensation principle according to which the unknown emf is opposed by another known emf till the two are balanced and no current flows through the galvanometer. With this method, problems of source instability can also be eliminated.

3. Units and dimensions

(a) Unit conversions

Q. 1. How many meters are there in i) 50 km, ii) 50 mm?

Ans.: i) 1 km = 1000 m, 50 km = 50000m = 5×10^4 m
ii) 1 mm = 10^{-3} m, 50 mm = 50×10^{-3} m = 0.05 m = 5×10^{-2} m

Q. 2. How many dm^3 are there in i) one m^3, ii) one cm^3, iii) one litre?

Ans.: i) 1m = 10 dm, $1m^3 = 10^3 \, dm^3$
ii) 1 cm = 10^{-2} m = 10^{-1} dm, $1 \, cm^3 = 10^{-3} \, dm^3$
iii) $1 \, L = 10^3 \, cm^3 = 1 \, dm^3$

Q. 3. Write freezing point and boiling point of water at one atmosphere pressure in three scales of temperature generally used.

Ans.: At one atmosphere of pressure, water freezes at 0°C, 32°F, and 273K, and boils at 100°C, 212°F, and 373.15K. Please note that the SI unit kelvin does not have a degree sign, therefore we write 273K rather than 273°K.

Q. 4. What are the conversion formulas for different temperature scales?

Ans.: Conversion from Fahrenheit scale to centigrade scale and vice versa can be done using the formula, $°F = 9/5 \, (°C) + 32$, $°C = 5/9 \, (°F - 32)$. For conversion from centigrade to kelvin scale 273.15 is added.

Q. 5. What are the conversion factors for the commonly used units of time?

Ans.: The SI unit of time is second but other units are also in common usage.
1 minute = 60 s, 1 hour = 60 minutes = 3600 s, 1 day = 24 hours = 1440 minutes = 86400 s.

Q. 6. What are the various conversion factors for units of energy?

Ans.: The SI unit of energy is joule but other units are also sometimes encountered. The conversion can be done as follows. 1 cal = 4.184 J, 1 eV = 1.602×10^{-19} J, 1 kWh = 3.6×10^6 J, 10^7 ergs = 1 J.

Q. 7. What are the conversion factors for commonly used units of pressure?

Ans.: The SI unit of pressure is Pascal. Conversion between various units can be done as follows. 1 atm = 101325 Pa = 760 torr ≈ 760 mm Hg, 1 bar = 10^5 Pa = 750.06 torr = 0.987 atm.

Q. 8. Define force. How can we convert the unit of force in C.G.S. system to that in SI system?

Ans.: Force is defined as the rate of change of momentum. For an unchanging mass, this is equivalent to mass x acceleration. The SI unit of force is Newton, N; 1N = 1 kg m s^{-2}. In C.G.S. system, the unit of force is Dyne (symbol: dyn); 1 dyn = 10^{-5} N.

Q. 9. List common conversion factors for different units of area, volume and density.

Ans.: Area: 10^6 mm² = 10^4 cm² = 1 m²
10^6 m² = 10^2 ha (hectare) = 1 km²
Volume: 10^6 mm³ = 10^3 cm³ = 1 L
10^6 cm³ = 10^3 L = 1 m³
Density: 10^3 kg/m³ = 10^3 g/L = 1 g/cm³

Q. 10. In unit conversions, how the number of significant figures in the final answer is determined?

Ans.: Conversion factors are not considered in the determination of significant figures in a calculation because these are assumed to be exact. For example, there are exact 60 seconds in a minute, 1000 m in a kilometer. However, conversion factors that come from measurements such as density or those which are approximations and have a limited number of significant figures should be considered in determining the significant figures in the final answer. Other rules are same as in any calculation.

Q. 11. A 100.0 ft x 9.0 ft lake has an average depth of 100.0 m. How many liters of water are there in the lake, if the lake is assumed to be full of water?

Ans.: Area = 100.0 x 9.0 = 900.0 ft². 1 ft = 12 inch = 12 x 2.54 = 30.48 cm. 1 ft² = 30.48 x 30.48 = 929 cm². Area = 900 x 929 = 836100 cm². Now 1m = 100 cm, Depth of the lake = 100 m = 100 x 100 = 10^4 cm. Volume of lake = Area x Depth = 836100 x 10^4 = 8.361 x 10^9 cm³. 1000 cm³ = 1L. Number of liters of water in the lake = 8.361 x 10^6. Since the number 9.0 used in the calculation has only 2 significant figures, the answer should be written as 8.4 x 10^6 L.

(b) Chemical conversions

Q. 1. A sodium chloride solution has a concentration 0.20 mole/dm³. Convert this concentration into molecules/cm³.

Ans.: 1 mole = 6.023 x 10^{23} molecules and 1 dm³ = 10^3 cm³. The concentration of sodium chloride solution = (0.20 x 6.023 x 10^{23})/10^3 = 1.2

x 10^{20} molecules/cm³. The number of significant figures in the final result depends on the precision of the starting information. The result is expressed in two significant figures since sodium chloride concentration has two significant figures.

Q. 2. How will you express the SI units of pressure (Pascal) and energy (Joule) in terms of SI base units?

Ans.: Pressure = Force/Area = Mass x Acceleration/ Area = kg m s^{-2} /m² = kg $m^{-1}s^{-2}$
Energy = Force x Distance = Mass x Acceleration x Distance = kg ms^{-2} m = kg m² s^{-2}

Q. 3. How many molecules of carbon dioxide are present in 4.4g of carbon dioxide?

Ans.: Molar mass of CO_2 = 44g. 4.4g of carbon dioxide = 4.4/44 = 0.10 mole = 0.10 x 6.023 x 10^{23} = 6.0 x 10^{22} molecules of carbon dioxide. The reported answer has two significant figures since the mass of carbon dioxide (4.4g) has two significant figures.

Q. 4. How many grams of sodium hydroxide are present in 10 moles of sodium hydroxide?

Ans.: The molar mass = 40 g/mol. 10 mole of sodium hydroxide = 400g. (The starting information, 10 moles is a count which is exact.)

Q. 5. How many kilograms of argon gas are present if there are 1000 atoms of argon?

Ans.: 1 mole = 6.023 x 10^{23} atoms of argon, 1 mole = molar mass of argon = 40g. 1 kg = 10^3g.
1000 atoms of argon = 1000/6.023 x 10^{23} moles = (1000/6.023 x 10^{23}) x 40g = (1000/6.023 x 10^{23}) x 40 x 10^{-3} kg = 6.64 x 10^{-23} kg argon gas. (1000 atoms is also a count which is exact.)

Q. 6. How many atoms of hydrogen are there in 100g of methane gas?

Ans.: Molar mass of methane = 16g = Mass of one mole of methane. 100g methane = 100/16 = 6.25 mole methane = 6.25 x 6.023 x 10^{23} = 37.64 x 10^{23} molecules of methane. One molecule of methane contains 4 hydrogen atoms. No. of hydrogen atoms in 100g of methane gas = 4 x 37.64 x 10^{23} = 150.56 x 10^{23} = 151 x 10^{23} atoms.

Q. 7. How many moles of mercury are present in a 25 mL sample of liquid mercury? Density of mercury = 13.5 g/cm³.

Ans.: Mass of mercury = density x Volume = 13.5 g/cm³ x 25 cm³ = 337.5 g. Molar mass of Hg = 200.59 g/mol. Number of moles of Hg = 337.5 /200.59 = 1.68. Since given data has two significant figures, the answer should be written as 1.7 moles.

Q. 8. How many equivalents of HCl are there in 100 mL of 0.01N HCl?

Ans.: Normality = Gram equivalents of solute/1L of solution. Thus Normality x Volume in milliliters = No. of milligram equivalents.
No. of equivalents of HCl = 0.01 x 100 x 10^{-3} = 1 x 10^{-3}. (Answer has only one significant figure since number 0.01 used in the calculation has one significant figure.)

Q. 9. How many moles of glucose are present in 1L of 8% (w/v) glucose solution?

Ans.: An 8% (w/v) glucose solution contains 8g glucose in 100 mL solution. One liter of glucose solution contains 80g glucose. Molar mass of glucose = 180. Number of moles of glucose in 1L of glucose solution = 80/180 = 0.4 moles. (Data as well as the answer has only one significant figure.)

Q. 10. A 5 liter container is filled with carbon dioxide gas at STP. How many moles of carbon dioxide are present in the container?

Ans.: One mole of an ideal gas occupies a volume of 22.4 L at STP (standard temperature and pressure). Number of moles of CO_2 = Volume of gas in liters/22.4 = 5/22.4 = 0.2. (Only one significant figure)

Q. 11. A 0.40M KCl solution contains 10 moles of KCl. Calculate the volume of solution.

Ans.: Number of moles of KCl in 1000 mL = 0.40. Therefore, 0.40 moles are present in 1L of solution. 10 moles will be present in (1 x 10)/0.4 = 25 L of solution. (Answer and data have two significant figures)

4. Preparation of solutions

Q. 1. Why measuring cylinder should not be used for preparing exact solutions?

Ans.: A measuring cylinder is a piece of laboratory glassware used to measure the volume of liquids. However, the graduation of measuring cylinder is only approximate and therefore, it cannot be used for quantitative analysis. Measuring flask, burette and pipette are properly calibrated and therefore, these glassware should only be used for measuring accurate volumes.

Q. 2. Describe how to prepare i) 250 mL of approximately 0.20 M NaOH solution and ii) 500 mL of approximately 0.50 N acetic acid solution?

Ans.: i) The number of grams of sodium hydroxide to be weighed = (Molecular weight of solute x Molarity x Total volume of solution to be prepared in mL)/1000 = (40 x 0.20 x 250)/1000 = 2.0 g. Transfer 2.0 g of sodium hydroxide into an appropriate container, dissolve completely in water, transfer to a measuring flask and make up the volume to 250 mL. Mix completely by inverting the stoppered flask several times and label it. The solution so prepared is only approximate since sodium hydroxide is not a primary standard.

ii) Glacial acetic acid available in the lab is usually approximately 17N. It is to be diluted to 0.50 N. The volume of acetic acid to be dissolved is calculated as follows using the relation, $N_1 V_1 = N_2 V_2$: 17 x V_1 = 0.50 x 500; V_1 = (0.50 x 500)/17 = 14.70 mL. 14.7 mL of glacial acetic acid, measured using a graduated cylinder, is transferred to a 500 mL measuring flask for further dilution to 500 mL mark. There is no need to measure the required volume of acetic acid with a pipette or burette since the solution so prepared is only approximate. The exact normality can only be determined by standardization of this solution.

Q. 3. What is a standard solution?

Ans.: A standard solution is a stable solution of accurately known concentration.

Q. 4. What are primary and secondary standard solutions?

Ans.: A primary standard solution is a solution of accurately known concentration prepared from a primary standard. A primary standard is a substance which is extremely pure and stable at room temperature. It should not be hygroscopic so that it can be weighed accurately and should be easily soluble in the solvent. Some examples of primary standards are oxalic acid ($C_2H_2O_4.2H_2O$), sodium carbonate (Na_2CO_3), potassium hydrogen phthalate ($KHC_8H_4O_4$), potassium dichromate ($K_2Cr_2O_7$).

The concentration of a secondary standard solution, on the other hand, is not very accurately known. The exact concentration of a secondary standard solution can usually be determined by titration with appropriate primary standard solution.

Q. 5. Why an acid such as hydrochloric acid or a base such as sodium hydroxide need to be standardized before use?

Ans.: Hydrochloric acid solution is usually prepared by dilution of the concentrated hydrochloric acid available in the lab. The concentrated acid is approximately 10N but the exact normality is not known and therefore, the solution prepared from this acid by dilution needs to be standardized.

Sodium hydroxide is available in the form of pallets which are weighed and dissolved in the solvent, usually water to prepare solution. However, solid sodium hydroxide is very hygroscopic, it absorbs moisture easily and therefore, it cannot be weighed exactly. For this reason it is not a primary standard and its solution should also be standardized before use.

Q. 6. How can we standardize sodium hydroxide and hydrochloric acid solutions?

Ans.: For acid-base titrations, oxalic acid (molecular weight 126 and equivalent weight 63) is usually used as primary standard. Standard oxalic acid solution is prepared by dissolving the appropriate quantity of accurately weighed oxalic acid in the solvent, usually water. The exact normality of sodium hydroxide solution is then determined by titrating it with standard oxalic acid solution using phenolphthalein as indicator.

For standardization of hydrochloric acid, a standard base is required which can be prepared as explained above. The exact normality of hydrochloric acid solution can be determined by titration with standard sodium hydroxide solution using phenolphthalein as indicator.

Q. 7. How will you prepare standard 0.1N acetic acid solution?

Ans.: Again acetic acid does not give primary standard solution. The concentrated glacial acetic acid available in the lab is approximately 17N. It is appropriately diluted and titrated with standard sodium hydroxide solution to know its exact normality. To prepare 250 mL of acetic acid solution, the required volume of acetic acid can be calculated using the relation, $N_1V_1 = N_2 V_2$: $17 \times V_1 = 0.10 \times 250$; $V_1 = (0.10 \times 250)/17 = 1.47$ mL. The acetic acid can be measured using a measuring (or graduated) cylinder since the solution is to be subsequently standardized. The volume of acetic acid taken should be little more than the calculated volume so that the prepared solution is slightly more concentrated than 0.1N which can then be accurately diluted to 0.1N using the normality equation, $N_1V_1 = N_2V_2$. In general, to prepare exact concentration of any solution which has to be subsequently standardized, it should be prepared at a slightly higher than the required concentration so that it can be appropriately diluted.

Q. 8. What is a stock solution? What are the advantages of using stock solutions?

Ans.: A stock solution is a concentrated solution which can be appropriately diluted to prepare working solutions. Working solution is one which is actually used for the experiment. There are a number of advantages of using stock solutions. Stock solutions, for example, can be prepared more accurately since larger amount has to be weighed; are generally more stable; reduce wastage of material and save time since multiple dilutions can be prepared from a single stock solution.

Q. 9. A 500mL stock solution of oxalic acid is prepared by dissolving 1.575g of oxalic acid in water. Five milliliter portion of this stock solution is diluted to 100 mL. Calculate the normality of final solution?

Ans.: Molecular weight of oxalic acid (including two water molecules) is 126 and the equivalent weight is 63.
Normality of concentrated stock solution = (1.575/63) x (1000/500) = 0.05
Normality of final diluted solution can be determined using the normality equation, $N_1V_1 = N_2V_2$.
Normality of diluted solution = (0.05 x 5) / 100 = 0.0025 = 2.500 x 10^{-3}.

Q. 10. How do you calculate dilution factor?

Ans.: The dilution factor is the ratio of concentration (normality) of concentrated solution and that of diluted solution. For example, volume of 0.5M stock solution required to prepare 100 mL of 0.1M solution is 20 mL, i.e., 5 times dilution is required; the dilution factor is 5. Similarly, a 1.0 N solution diluted 5 times will be 0.02 N; a 0.02 N solution further diluted 10 times will be 0.002N.
　　Dilution factor can also be defined as the final volume divided by the initial volume. For example if 0.1 mL of a stock solution is diluted to 10 mL, the dilution factor is 10/0.1= 100.

Q. 11. What volume of 0.8N stock solution is required to make 100 mL of 0.2N solution?

Ans.: This requires 4 times dilution and therefore, 25 mL of stock solution is to be diluted to 100 mL. The same can also be calculated using the normality equation, $N_1V_1 = N_2V_2$ (0.8 x V_1 = 0.2 x 100, V_1 = 25 mL).

Q. 12. What is serial dilution?

Ans.: Serial dilution is a step-wise dilution and the dilution factor is generally constant for various steps. For example, a ten-fold serial dilution means a 1N stock solution is diluted 10 times in each step resulting in solutions which are 0.1N, 0.01N, 0.001N and so on.

Q. 13. What concentration units are commonly used for the preparation of reagent solutions in laboratory?

Ans.: Reagent solutions are usually prepared in the following concentration units: molarity (M), normality (N). molality (m), mole fraction (x) and percent (%).

Q. 14. Define Molarity, molality, normality, mole fraction and percentage.

Ans.: Molarity, molality, normality, mole fraction and percentage are different ways of expressing concentration of solution.

Molarity is the number of moles of solute present in one liter of solution (Moles of solute/1L of solution). Normality is the number of gram equivalents of solute present per liter (dm^3) of solution at any given temperature (Gram equivalents of solute/1L of solution). Molarity is the most commonly used unit of concentration since normality changes with gram equivalent weight which depends on the role of the reactive species in the reaction. Let us take the example of sulphuric acid. In acid-base reactions where hydrogen ion is the reactive species, 1M sulfuric acid would be 2N whereas in precipitation reactions where sulphate ion is the reactive species, 1M H_2SO_4 solution = 1N. Similarly in ferric thiosulfate, $Fe_2(S_2O_3)_3$, if the reactive species is Fe, a 1M solution would be 2N (2 Fe atoms are present). However, if thiosulfate is the active species, a 1M solution would be 3N because 3 moles of thiosulphate ions are present per mole of $Fe_2(S_2O_3)_3$. In redox reactions normality refers to the number of electrons donated or accepted per mole of reactant. Thus molarity describes the number of moles of complete substance per liter of solution whereas normality describes only the moles of reactive species per liter of solution. Gram equivalent weight is a measure of the reactive capacity of molecule. Normality is always a multiple of molarity.

Molality is the number of moles of solute dissolved in 1kg of solvent (Moles of solute/1 kg of solvent). Mole fraction is another way of expressing the concentration of solution. It is defined as the number of moles of a given component divided by the total number of moles present. Mole fraction has no units since it is a fraction. The sum of the mole fractions of all the components present in a system is always unity. Another related term used is mole percent which is percentage of the total number of moles present. Mole percent is equal to the mole fraction multiplied by 100. The sum of the mole percent of all the components present in a system is 100. Similarly mass fraction is the ratio of the mass of a given component divided by the total mass. It can be calculated for any compound using empirical or molecular formula.

Percentage can be defined in different ways such as weight per unit weight (weight of solute divided by weight of solution and multiplied by 100; units: g/100g), weight per unit volume (weight of solute divided by the volume of solution and multiplied by 100; units: g/100 mL) and volume per unit volume (volume of solute divided by total volume of solution and multiplied by 100; units: mL/100 mL). Weight per unit volume is generally used for dissolution of a solid solute in liquid solvent and volume per unit volume for the dissolution of a liquid solute in liquid solvent. Percentage units are useful if the molecular weight of solute is not known.

Some other ways to express concentration of a chemical solution are parts per million parts (ppm) and parts per billion parts (ppb). These units, defined for extremely dilute solutions, are relatively less frequently used. For aqueous solutions, the density of water is commonly assumed to be 1.00 g/mL and therefore, 1ppm corresponds to 1mg/L and 1 ppb corresponds to 1 µg/L.

Q. 15. How can we convert a percent solution (w/v) to molarity and a molar solution to percent?

Ans.: To convert from percent solution to molarity, multiply the percentage of solution by 10 to get g/L and then divide by formula weight. For example, a 9.0% solution of KNO_3 (Formula weight = 101) will be 90/101 = 0.89 M.

To convert from molarity to percent, multiply the molarity value by formula weight to convert moles to grams and divide by 10 to change to per 100 mL. For example, a 0.50M NaCl (Formula weight = 58.5) solution will be (0.50 x 58.5)/10 = 2.9 % (w/v).

Q. 16. Why reagent solutions should always be kept covered with stopper?

Ans.: Reagent solutions should always be kept covered with stopper to avoid errors caused by evaporation of solvent. Evaporation of solute is also possible if it is volatile. Moreover, some reagents such as sodium hydroxide can react with atmospheric carbon dioxide causing error.

Q. 17. How many moles/grams of KCl are contained in 10 mL of 5.0M KCl solution?

Ans.: 5.0 moles of KCl are present per liter of solution. No. of moles of KCl in 10mL of solution = (5.0/1000) x 10 = 0.050.
The number of grams of KCl in one mole = 74.5 (molar mass of KCl). No. of grams of KCl in 10mL of solution = Number of moles of KCl in 10 mL x molar mass of KCl = 0.050 x 74.5 = 3.7.

Q. 18. Describe how you will prepare an exact solution if the given sample is not 100% pure?

Ans.: For example, we have to prepare exact 0.10M solution of KCl when the salt is only 90% pure. We know that the mass of solute required for preparation of solution is calculated as (Molecular weight of solute x Molarity x Total volume of solution to be prepared in mL)/1000, when the solute is assumed to be 100% pure. If the solute is not 100% pure, larger amount has to be dissolved to get the desired concentration. So the calculated mass is to be divided by the percentage purity of the sample. The mass of KCl required = (74.5 x 0.10 x 500)/(1000 x 0.90) = 4.14g.

Q. 19. You are provided with a sample of phosphoric acid which is 85% pure. Describe how you would prepare 250 mL of exact 0.50 M phosphoric acid solution?

Ans.: The gram formula weight of phosphoric acid is 98.00 g/mol and the density of 85.0% phosphoric acid is 1.68 g/mL. To calculate the volume of 85% phosphoric acid required, we have to divide the calculated mass by the purity as well as density of sample. The volume of 85% phosphoric acid required for the preparation of 250 mL of exact 0.50 M phosphoric acid solution = (98.0 x 0.50 x 250) / (1000 x 0.85 x 1.68) = 8.6 mL.

Q. 20. Explain the conversion of molar solution to ppm and vice versa.

Ans.: Parts per million parts (ppm) refers to the parts of solute per million parts of solution. The general formula for preparation of solution in ppm units can be written as

$$ppm = \frac{Weight\ or\ volume\ of\ solute}{Weight\ or\ volume\ of\ solution} \times 10^6$$

In general, for dilute aqueous solutions one part per million parts (ppm) is considered equal to 1 mg/L since the density of water can be taken as approximately unity. Following examples demonstrate the conversion of molar solution to ppm and vice versa.

i) Conversion of 0.01M solution of HCl to ppm units: A 0.01M solution of HCl contains 0.01 moles = 0.01 x 36.5 = 0.365g = 365 mg of HCl per liter of solution. So the concentration of HCl is 365 ppm.

ii) Conversion of 100 ppm NaCl solution to molarity: A 100 ppm NaCl solution contains approximately 100 mg of NaCl = 100/58.5 = 1.71 mmoles = 1.71 x 10^{-3} moles per liter of solution. So the NaCl solution is 1.71 x 10^{-3} M or 1.71 mM.

5. Distribution law

Q. 1. State Nernst distribution law.

Ans.: Nernst distribution law is based on a generalization which relates to the distribution of a solute between two immiscible solvents. It states 'that at equilibrium, a solute distributes itself between two immiscible solvents in such a way that the ratio of its concentration in the two solvents is always a constant at a given temperature'. The conditions for validity of distribution law are i) the solute should not react with any of the solvents and ii) it should be present in the same molecular state in the two solvents i.e. it should neither dissociate nor associate in any of the solvents.

Q. 2. Define partition coefficient/distribution coefficient. Can it be called equilibrium constant?

Ans.: If a solute is distributed between two immiscible solvents, at equilibrium the ratio of the concentration of solute in the two solvents at a given temperature is defined as the partition or distribution coefficient of the solute. Mathematically, partition/distribution coefficient, represented by symbol P or $K_D = c_2/c_1$, where c is the concentration of solute and subscripts 1 and 2 refer to the two immiscible solvents in which the solute is distributed. Since partition coefficients can vary over a very wide range, it is also frequently represented in logarithmic form (log P).

Yes, it can be called equilibrium constant because partition equilibrium is a special case of chemical equilibrium. The equilibrium reaction involved in this case is the distribution of solute in two immiscible solvents.

Q. 3. For the partition of a solute between an aqueous and organic solvent, partition coefficient should be written as $[c_{org}]/[c_{aq}]$ or $[c_{aq}]/[c_{org}]$?

Ans.: There is no established convention with regard to whether $[c_{org}]$ or $[c_{aq}]$ should be in numerator in the expression for partition coefficient. Therefore, one should always specify the way it is calculated. In the literature, however the more commonly encountered definition is $[c_{org}]/[c_{aq}]$ and therefore, a substance with high partition coefficient is considered to have high lipid solubility.

Q. 4. A solid solute X is distributed between water and benzene at constant temperature. At equilibrium 10 mL of benzene layer was found to contain 0.75g of solute while 100 mL of water layer contained 0.50g of solute. What is the distribution coefficient of solute in water and benzene?

Ans.: Distribution coefficient, $K_D = c_2/c_1$
Concentration of solute in benzene layer, $c_2 = 0.75/10 = 0.075$ g/mL
Concentration of solute in water layer, $c_1 = 0.50/100 = 0.005$ g/mL
Distribution coefficient, $K_D = 0.075/0.005 = 15.0$
Please note that distribution coefficient, being the ratio of two concentrations is dimensionless. Also for calculation of distribution coefficient, concentrations can be expressed in any convenient unit but both the concentrations should be in the same units.

Q. 5. Two (2.0) grams of iodine are shaken with a mixture of 75 mL of water and 25 mL of carbon tetrachloride at constant temperature. What will be the concentration of iodine in the two layers at equilibrium? Distribution coefficient of iodine between carbon tetrachloride and water $c_{CCl_4}/c_{H_2O} = 85$.

Ans.: The concentration of solute in the two layers can be calculated if distribution coefficient is known. Let the amount of iodine present in the CCl_4 layer be x g. The water layer will then contain $2 - x$ gram of iodine.

The concentrations of iodine in the CCl_4 and water layer in g/mL will be $x/25$ and $(2-x)/75$, respectively. The distribution coefficient,

$$\frac{[c_{CCl_4}]}{[c_{H_2O}]} = \frac{x/25}{(2-x)/75} = 85.$$

The value of x, calculated from this expression is 1.93. Thus CCl_4 layer contains 1.93g iodine and water layer contains $2 - 1.93 = 0.07$ g iodine. Hence the concentration of iodine in CCl_4 layer $= 1.93/25 = 0.077 = 7.7 \times 10^{-2}$ and the concentration of iodine in water layer $= 0.07/75 = 9.3 \times 10^{-4}$ g/mL.

Q. 6. What are the limitations/conditions for validity of distribution law?

Ans.: i) The law is valid only at constant temperature. Thus the use of a thermostat is required. ii) The solute should be present in the same molecular state in the two solvents i.e. it should neither dissociate nor associate in any of the solvents and it should not react with any of the solvents. iii) The law is valid only at equilibrium. So the concentration of solute should be determined only when the equilibrium has been established. iv) The concentration of solute in the two solvents should be low since the law is applicable only in dilute solutions. Alternatively the values can be extrapolated to zero concentration. v) The two solvents should be immiscible or only slightly miscible. Also the extent of mutual solubility of the solvents should remain unaltered by the addition of solute to them.

Q. 7. How distribution law is modified when the solute associates in one of the solvents?

Ans.: If, for example, solute (S) remains unassociated in the aqueous solvent 1 and n molecules of solute associate in the organic solvent 2, then the equilibrium $nS\,(aq) \rightleftharpoons S_n\,(org)$ exists so that the partition/distribution coefficient, $K_D = [c_{org}]/[c_{aq}]^n$ or $K_D = [c_2]/[c_1]^n$. We can also determine n, the number of molecules which combine to form the associated molecule. If we take logarithm on both sides of relation, $K_D = [c_2]/[c_1]^n$, we can write

$\log [c_2] = n \log [c_1] + \log K_D$. Thus a plot of $\log [c_2]$ (concentration in organic layer) against $\log [c_1]$ (concentration in aqueous layer) gives a straight line with slope n and antilog of intercept gives partition/distribution coefficient, K_D.

Q. 8. How distribution law is modified when the solute dissociates in one of the solvents?

Ans.: If a fraction of the solute dissociates in one of the solvents, the partition coefficient will be equal to the ratio of the concentrations of undissociated solute in the two solvents. If for example, the solute dissociates in solvent 2 and α is the degree of dissociation, the ratio of the concentration of undissociated solute in the two solvents will be $c_1/\{c_2(1-\alpha)\}$. The distribution coefficient, K_D can thus be defined as $K_D = c_1/\{c_2(1-\alpha)\}$. Degree of dissociation (α), defined as the fraction of the total solute which dissociates, can also be calculated from this expression if K_D is known.

Q. 9. What are the major applications of distribution law?

Ans.: i) Distribution law can be used for the determination of degree of association or degree of dissociation of a solute as discussed above in Qs. 7 & 8, respectively. ii) Distribution law is also useful in determining the solubility of a solute in a solvent. When a solute is shaken with two immiscible solvents, at equilibrium both the solvents are saturated with the solute. Since the solubility also represents concentration of saturated solution, we can write the distribution law as $c_2/c_1 = S_2/S_1 = K_D$, where S_2 and S_1 are the solubilities of the solute in the two solvents. Hence knowing the value of the Distribution coefficient (K_D) and the solubility of solute in one of the solvents, the solubility of solute in the second solvent can be calculated. For example, the solubility of iodine in water is 0.30 g/L. If the distribution constant of iodine in CCl_4 and water is 100, the solubility of iodine in carbon tetrachloride will be 30g/L. iii) Another major application of distribution law is in the extraction of a solute dissolved in one solvent by another solvent, the two solvents being immiscible or only slightly miscible. For example, solvent extraction is frequently employed in the extraction of a dissolved organic substance from aqueous solution with solvents such as benzene, chloroform, carbon tetrachloride etc. in the laboratory or industry. It can be shown that the amount left unextracted = $W [K_D V/(K_D V + v_1)]^n$, where W = Initial amount present in solution, K_D = Distribution coefficient expressed as $[c_{aq}]/[c_{org}]$, V = Volume of solution, v_1 = Volume of extracting solvent, n = Number of extractions performed. Thus percentage recovery is always more when extraction is performed in a number of steps.

Q. 10. At 25^0C the distribution coefficient of H_2S between water and benzene, defined as C_{aq}/C_{org} = 0.20. What is the minimum volume of benzene necessary to extract in a single step 90% of H_2S from 1L of a 0.1M aqueous solution of H_2S?

Ans.: Using distribution law we can easily calculate the volume of extracting solvent required to extract a given amount of solute. We know that the amount of solute left unextracted = $W [K_D V/(K_D V + v_1)]^n$, where W = Initial amount present in solution, K_D = Distribution coefficient, V = Volume of solution, v_1 = Volume of extracting solvent, n = Number of extractions performed. For 90% extraction, if the initial amount of solute W is taken as 100, the amount left unextracted will be 10. The volume of solution, V = 1L. Since extraction is to be performed in a single step, n =1. Therefore, $10 = 100 [0.20 \times 1/(0.20 \times 1 + v_1)]^1$ which can be solved to get v_1 = 1.8L. Thus minimum volume of benzene required for 90% extraction = 1.8L.

Q. 11. Which precautions should be observed during experimental determination of distribution coefficient?

Ans.: i) Although the two solvents are considered to be immiscible, they do not have zero solubility in each other and dissolution of one liquid in the other will cause error. Therefore, before starting the experiment, the two liquids should be shaken together thoroughly to get a saturated solution of one solvent in another. ii) The solute should preferably be first dissolved in the solvent in which it has higher solubility and then shaken with the other solvent. iii) The mixture should be shaken thoroughly for sufficient time so that the equilibrium is attained. iv) After equilibrium has been attained, the mixture should be allowed to stand for enough time for clear separation of the two layers. v) While withdrawing aliquot, due care is needed to make sure that one layer is not contaminated with the other. vi) The distribution of solute and the distribution coefficient are temperature-dependent and therefore, the assembly must be kept in a thermostat so that the temperature remains constant during the experiment.

Q. 12. For an unknown system, how can we find out that the solute does not exist in the same molecular state in the two solvents?

Ans.: The experiment is performed using different initial concentrations of solute. If the distribution coefficient, calculated as the ratio of concentration of solute in the two solvents at equilibrium, does not remain constant, it is likely that the solute exists in different molecular states in the two solvents. Please note that the distribution coefficient should remain constant at constant temperature irrespective of the initial concentration of solute.

Q. 13. Why water or aqueous potassium iodide solution is added during titration of non-aqueous layer in the experimental determination of partition coefficient of iodine between water and a non-aqueous liquid?

Ans.: Since aqueous sodium thiosulfate is used for the titration of both aqueous and non-aqueous layers, non-aqueous layer is not miscible with the titrant. When water/KI solution is added to the flask, iodine gets gradually transferred to the water/KI layer from where it is titrated with thiosulfate.

Q. 14. What are the factors on which distribution coefficient depends?

Ans.: For a given pair of solvents, distribution coefficient of a solute depends only on temperature.

Q. 15. For the titration of iodine with potassium thiosulphate, why starch is added only when the color of the solution is pale yellow?

Ans.: In iodometry, iodine is titrated with thiosulfate and starch is used as indicator. However, starch should be added near the end point, when the solution color is pale yellow because starch reacts with iodine to form a dark blue complex. If it is added earlier, then some of the iodine tends to remain bound to starch particles, even after the equivalence point is reached, thus producing an unwanted error.

Q. 16. Why freshly prepared starch solution is needed for iodometric titrations?

Ans.: Freshly prepared starch solution should be used because with time starch undergoes photochemical degradation and forms a colloidal solution which should not be used for the titration.

Q. 17. Write reaction between iodine and sodium thiosulphate.

Ans.: The reaction between iodine and sodium thiosulfate is a redox reaction in which iodine is reduced to iodide ions by thiosulfate

$$I_2 (s) + 2Na_2S_2O_3 (aq) \rightarrow Na_2S_4O_6 (aq) + 2NaI (aq)$$

Q. 18. Can partition/distribution coefficient be used to calculate the standard free energy change of a chemical reaction?

Ans.: Since partition/distribution coefficient is an equilibrium constant for the distribution of solute in two immiscible solvents, it is related to standard free energy change by the expression $\Delta G^0 = -RT \ln K_D$ or $\Delta G^0 = -RT \ln P$.

Q. 19. How do you define mathematically the distribution coefficient for the distribution of benzoic acid between water and benzene?

Ans.: Benzoic acid remains as such in water. However, when dissolved in benzene, benzoic acid associates to form dimers. Thus modified form of distribution law is applicable. Accordingly, $K_D = [c_{org}]/[c_{aq}]^2$ or $K_D = [c_{org}]^{1/2}/[c_{aq}]$

Q. 20. What is the principle of the method used for the determination of equilibrium constant for the reaction, $KI + I_2 \rightarrow KI_3$ using distribution coefficient method? How equilibrium constant is related to the standard free energy change for the reaction?

Ans.: When iodine distributes between aqueous potassium iodide solution and an organic solvent, the following two equilibria are established together. i) $I_2 + I' \rightleftharpoons I_3'$ in the aqueous layer due to the ionic species and I_2 (aq) $\rightleftharpoons I_2$ (org) due to the partition phenomenon itself. The equilibrium constant for the reaction, is given by the following expression.

$$K_{eq} = [I_3']_{eq}/([I_2]_{eq} \times [I']_{eq})\ L mol^{-1}$$

$[I_2]_{aq}$ = Total I_2 concentration in the aqueous layer = $[I_2]_{eq} + [I'_3]_{eq}$ since in aqueous layer at equilibrium, iodine is present in two forms, $[I_2]$ and $[I_3']$.

$[I_2]_{org}$ = I_2 concentration in organic layer

K_D = Distribution coefficient of iodine = $[I_2]_{org}/[I_2]_{eq}$ because distribution law applies only to similar molecular species, i.e. $[I_2]$ in the two layers. So

$[I_2]_{eq} = [I_2]_{org}/K_D$, $[I_3']_{eq} = [I_2]_{aq} - [I_2]_{eq}$ and $[I']_{eq} = [I']_0 - [I_3']_{eq}$

$[I'_0]$ = Initial KI concentration

Thus experimental determination of concentration of iodine in the aqueous and organic layers and knowledge of the partition coefficient of iodine between water and organic solvent, the equilibrium constant for the reaction can be determined.

6. Phase rule

Q. 1. Write phase rule equation and explain the terms involved.

Ans.: According to phase rule equation, $F = C - P + 2$ for a system at equilibrium. F = Number of degrees of freedom. Degree of freedom is the number of independent intensive variables which are required to define the system. C = Number of components. Number of components is the minimum number of independent species necessary to define the composition of all the phases present in the system. P = Number of phases present at equilibrium. A phase is a part of the system which is homogeneous in chemical composition and physical state. The numerical value 2 usually signifies the two degrees of freedom of temperature and pressure.

Q. 2. What is the criterion for phase equilibrium?

Ans.: Phase equilibrium is a state of balance between different phases present in the system. Balance means that the two phases in equilibrium must have the same temperature, pressure and chemical potential so that there is no net transfer of heat, work and chemical energy from one phase to another.

Q. 3. Describe the general procedure used to determine the number of components in a chemically reactive system.

Ans.: The number of components is the minimum number of independently variable constituents that could be used to define each phase individually. In order to calculate the number of components in any chemically reactive system, the relation, C = s − r is used where s is the total number of different chemical species present and r is the number of restrictions imposed on the independent variation of these constituents such as the reaction equilibria involved, requirement of electroneutrality in a phase containing ions, the relations between concentrations due to initial conditions etc. These restrictions are other than those required for phase equilibrium. Let us take the example of sublimation of solid ammonium chloride in a closed container. The sublimation reaction can be written as $NH_4Cl\ (s) \rightleftharpoons NH_3\ (g) + HCl\ (g)$. In this system the total number of chemical species (s) is 3 and the number of restrictions imposed (r) is 2. One is the reaction equilibrium condition and the other is the fact that dissociation of NH_4Cl always produces equivalent amounts of NH_3 and HCl. Since products consist of a single gaseous phase we need to know the concentration of only one of these. Thus it is a one component system (C = s − r = 3 − 2 = 1). The dissociation of $CaCl_2$; $CaCl_2(s) \rightleftharpoons CaO(s) + CO_2(g)$, on the other hand is a two component system since r = 1. Here also the products are formed in equivalent amounts but they constitute two different phases and therefore the composition of each phase has to be described individually.

Q. 4. Liquid water is in equilibrium with water vapour in a closed container. How many degrees of freedom are there in the system?

Ans.: According to phase rule equation, F = C − P + 2 for a system at equilibrium. In this system, C = 1, P = 2. So the number of degrees of freedom, F = 1. This means only one variable needs to be defined to define the state of the system. Say if we define temperature, vapour pressure of water is automatically fixed and if we define vapour pressure, temperature is automatically fixed.

Q. 5. What are condensed systems? Write phase rule equation for condensed systems.

Ans.: Systems which contain only solid and liquid phases and no gaseous phase are called condensed systems. Since condensed systems are little affected by changes in pressure, one degree of freedom can be discarded. So for such systems, phase rule equation takes the form, $F = C - P + 1$. This is called condensed or reduced phase rule equation.

Q. 6. Ice is in equilibrium with liquid water in a closed container. What is the number of degrees of freedom in the system?

Ans.: Ice in equilibrium with liquid water is a condensed system since no gaseous phase is present. According to the condensed phase rule equation, $F = C - P + 1 = 1 - 2 + 1 = 0$. The number of degrees of freedom, $F = 0$ means that the system is invariant. In an invariant system no variable needs to be defined to define the state of the system. Thus ice is in equilibrium with liquid water at a fixed temperature which is the melting point of ice (0^oC).

Q. 7. What are the maximum number of phases which can coexist and the maximum number of degrees of freedom in i) a one-component system and ii) a two-component condensed system?

Ans.: For a one-component system phase rule equation, $F = C - P + 2$ can be written as $F = 3 - P$. For a two-component condensed system also $F = 3 - P$ since the condensed phase rule equation, $F = C - P + 1$ is applicable. Since F, the number of degrees of freedom cannot be negative the maximum number of phases which can coexist in a one-component system and a two-component condensed system is three. The minimum number of phases which can exist in any system is one and therefore, the maximum number of degrees of freedom is two.

Q. 8. What are the simplest types of 2-component systems, which can be easily studied in the lab? Give examples.

Ans.: For a 2-component system, three variables, temperature, pressure and composition have to be defined to define the state of the system. This means that the phase diagram should be 3-dimensional. Since condensed systems are little affected by changes in pressure, a considerable simplification is achieved by assuming that pressure remains constant and then 2-dimensional temperature versus composition diagrams can be used. Thus the simplest type of 2-component systems are two-component solid-liquid condensed systems in which the two components are completely miscible with one another in the liquid state, they do not form any compound and on solidification give only an intimate mixture, known as eutectic. These systems are also called simple eutectic systems. Some examples are lead-silver, lead-antimony, naphthalene - diphenylamine and naphthalene-p-toluidine. Due to the convenience of the temperature range involved, naphthalene- diphenylamine and naphthalene-p-toluidine systems are most commonly studied in the lab.

Q. 9. What is a phase diagram?

Ans.: A phase diagram is a graphical representation of the conditions (temperature, pressure and composition) at which thermodynamically distinct phases occur and coexist at equilibrium. For 1-component systems, the two variables chosen are pressure and temperature since composition or concentration in all phases is invariably 100 percent. They are therefore, represented by pressure versus temperature phase diagrams. Since two component condensed solid - liquid systems are little affected by changes in pressure, temperature versus composition phase diagrams are used. From the phase diagram physical state of a system may be deduced once temperature, pressure and composition are defined.

Q. 10. How phase diagrams of two-component solid-liquid condensed systems are determined?

Ans.: The phase diagram is a plot of freezing points of mixtures of different composition against composition. The freezing points of various mixtures, determined by thermal analysis, are plotted against composition.

Q. 11. What is thermal analysis and how cooling curves are determined?

Ans.: The shapes of freezing point curves can be obtained experimentally by thermal analysis. A mixture of known composition is heated to a high enough temperature so that it is homogeneous. Then it is allowed to cool at a regulated rate. The temperature is plotted as a function of time. The resulting curves are called cooling curves and these curves are used to obtain freezing point data for the construction of phase diagrams.

Q. 12. In the experimental determination of cooling curves, why is it necessary to place the inner tube containing mixture in an outer jacket?

Ans.: When the inner tube containing mixture is placed in an outer tube, the outer tube serves as an air jacket which ensures uniform cooling of the mixture in the inner tube.

Q. 13. Explain cooling curves of a single component system (pure compound) and a 2-component solid-liquid system.

Ans.: In a single component system on cooling liquid melt, temperature shows a regular fall till solidification starts. As long as the process of solidification continues the temperature remains constant. A small dip in the curve is obtained only when super cooling occurs. The constancy of temperature during crystallization (or melting) is an important practical criterion of purity of a substance. After complete solidification, temperature of the solid falls. This can also be explained on the basis of the condensed phase rule equation, $F = C - P + 1$. When $C = 1$, $F = 2 - P$. So temperature changes (decreases) when only liquid or only solid phase is present ($P = 1$, $F = 1$). When solid and liquid phases are in equilibrium in the middle horizontal region, there are two phases ($P = 2$) and the system is invariant, $F = 0$. In an invariant system no variable needs to be defined to define the state of the system and therefore, temperature remains constant during crystallization.

Now let us take the case of a system consisting of two components, A and B which are completely miscible with each other in the molten state and they do not form any compound. When a homogeneous mixture consisting of both the components in molten state is cooled, it cools in a regular manner till the solution reaches a saturation point with respect to one of the components, say B. Then component B precipitates out of the solution and this releases the latent heat of fusion, the rate of cooling slows down and a break in the curve appears at a temperature, T_f which represents the freezing point of the mixture. As the solution is further cooled, component B continues to crystallize out till the solution reaches saturation with respect to component A also. Further cooling results in precipitation of component A as well. Now the system consists of three phases, solid component A, solid component B and solution in equilibrium. At this stage the system becomes invariant and therefore, temperature remains constant as long as three phases are in equilibrium. When all the liquid has solidified, only two phases are present and the temperature falls again. The behavior of two-component system discussed above can also be explained on the basis of the condensed phase rule equation, $F = C - P + 1$. When $C = 2$, $F = 3 - P$. In the first segment of cooling curve, a single liquid phase is present, $F = 2$, temperature falls. When component B starts crystallizing, there are two phases, $F = 1$, temperature can fall further. However, when both components A and B start crystallizing, the system has three phases in equilibrium, $F = 0$, which means that as long as three phases are in equilibrium, temperature has to remain constant and the composition of the system does not change. The heat loss is compensated by the heat gain due to latent heat of crystallization of the sample. The horizontal plateau in the curve is called eutectic halt. The temperature at which eutectic halt occurs is called eutectic temperature and the corresponding composition of liquid is called eutectic composition.

Q. 14. What are simple eutectic systems?

Ans.: Simple eutectic systems are two-component solid-liquid condensed systems in which the two components are completely miscible with one another in the liquid state; they do not form any compound and on solidification give only an intimate mixture, known as eutectic.

Q. 15. What are liquidus and solidus curves?

Ans.: Liquidus curve is a line in the phase diagram above which only liquid is present or the line above which no solid exists. It separates the liquid and liquid + solid regions in the phase diagram. Solidus curve is the line below which only solid is present or line below which no liquid exists. Solidus line separates the liquid + solid and solid regions in the phase diagram. The intermediate region between the liquidus and solidus lines is the two phase region where liquid and solid phases coexist.

Q. 16. Explain general behaviour of the phase diagram of a 2-component simple eutectic solid-liquid system.

Ans.: Cooling curves are used to determine the freezing points of pure components as well as mixtures of the two components with compositions varying from 100% component A to 100% component B. The phase diagram is then obtained by plotting the freezing points of the mixtures against their composition. In the first portion of the plot, when increasing amounts of B are added to pure A, solid A starts separating out; solid A and liquid solution are in equilibrium and freezing points decrease. This continues up to a certain composition beyond which the freezing points increase. In the second portion of the plot solid B and liquid solution are in equilibrium. The two curves meet at the eutectic point where all the three phases, solid A, solid B and liquid solution are in equilibrium. Since for a two-component condensed system phase rule equation is written as $F = 3 - P$, the system is invariant at the eutectic point. Eutectic temperature and eutectic composition can be determined from the intersection of the two freezing point curves. Below eutectic temperature only solid phase consisting of solids A and B exists, there is no liquid phase.

Q. 17. Define eutectic temperature and eutectic composition.

Ans.: Eutectic point is a characteristic point in the phase diagram where the two liquidus curves meet the solidus curve. The eutectic point is the only point on the diagram where this is true. At this point the two components crystallize out simultaneously in the same proportion in which they are present in the eutectic liquid. It has a lower freezing point than either of the components or their mixtures in any proportion. This temperature is called eutectic temperature and the composition corresponding to this point is called eutectic composition.

Q. 18. Why does a mixture corresponding to eutectic composition behave like a pure substance?

Ans.: The eutectic mixture is that mixture which has the lowest freezing point. At this point the two components crystallize out simultaneously in the same proportion in which they are present in the eutectic liquid. The freezing temperature therefore remains constant and solidification is completed at constant temperature. Thus it behaves like a pure substance.

Q. 19. What is the difference between triple point and eutectic point?

Ans.: Triple point is a point in a 1-component system where all the three phases are in equilibrium while eutectic point is a point in a two-component solid-liquid simple eutectic system where three phases are in equilibrium. The common feature is that at both these points, the system is invariant (F=0) and three phases are in equilibrium.

Q. 20. What are congruent and incongruent melting points?

Ans.: Congruent and incongruent melting points refer to systems in which the two components form a stable compound in the liquid state. A compound which melts sharply at a constant temperature into a liquid of the same composition as the solid is said to possess a congruent melting point. If the compound formed by the combination of two components decomposes on heating giving a new solid phase and a solution with a composition different from that of the original solid, it has an incongruent melting point.

Q. 21. What is peritectic point?

Ans.: Peritectic point is a point on a phase diagram where a reaction takes place between a previously precipitated phase and the liquid to produce a new solid phase. When this point is reached, the temperature must remain constant until the reaction has run to completion. A peritectic point is also an invariant point.

7. Thermochemistry

Q. 1. What is the difference between isothermal and adiabatic calorimeter?

Ans.: A calorimeter is used to determine the enthalpy change in a reaction. In an isothermal calorimeter temperature remains constant and therefore change in some other associated property of the system caused by the absorption or evolution of heat is measured. An adiabatic calorimeter is isolated from the surroundings and therefore, heat absorbed/evolved lowers/raises the temperature of the system which is measured. Adiabatic calorimeters are more commonly used in the lab due to their simplicity.

Q. 2. What kind of calorimeter is thermos flask?

Ans.: A thermos flask is thermally insulated from the surroundings and therefore, it can be conveniently employed as an adiabatic calorimeter in the lab for routine measurements.

Q. 3. What do you understand by the water equivalent or heat capacity of a calorimeter?

Ans.: If a reaction is carried out in a calorimeter, some heat will also be taken up by the calorimeter. The heat capacity of the calorimeter is the amount of heat required to raise the temperature of the calorimeter by 1^0C. The heat capacity of a calorimeter is also called water equivalent because it is also equivalent to the mass of water having the same heat capacity as the calorimeter.

Q. 4. Why it is necessary to determine the water equivalent/heat capacity of a calorimeter?

Ans.: As a result of a chemical reaction, some heat is evolved or absorbed. Since the reaction is carried out in the calorimeter, the calorimeter vessel must also be in thermal equilibrium with the reaction mixture, that is the loss or gain of heat by the calorimeter vessel should also be taken into account. Water equivalent/heat capacity is a measure of the amount of heat gained or lost by the calorimeter per degree change in temperature.

Q. 5. What principle is used for the estimation of water equivalent of a calorimeter?

Ans.: For the determination of water equivalent (W) of a calorimeter, the basic principle involved is that the total heat lost is equal to the total heat gained. The method used involves mixing of hot water and cold water. Heat will be lost by hot water and gained by cold water as well as calorimeter. From the volume of hot and cold water (say 100 mL each), temperature of hot water (t_h) and cold water (t_c) at the time of mixing and the highest temperature attained after mixing (t_m), the water equivalent of calorimeter can be determined as follows.

Heat lost by hot water = Heat gained by cold water + Heat gained by calorimeter.

$$100\,(t_h - t_m) = 100\,(t_m - t_c) + W(t_m - t_c) = (100 + W)\,(t_m - t_c)$$

$$W = \left[\frac{100\,(t_h - t_m)}{(t_m - t_c)} - 100\right]$$

Water equivalent/heat capacity (W) of a calorimeter can also be determined by first using a salt of known enthalpy of reaction. The enthalpy of reaction of unknown system can then be determined, using the calculated value of W.

Q. 6. What is the unit of the water equivalent/heat capacity of a calorimeter?

Ans.: In CGS system, the unit of heat capacity of a calorimeter is cal/°C while in SI units it is J/K. The value calculated using expression derived in Q. 5 will be in cal/°C. In order to obtain the value in SI units, it is to be multiplied by 4.184.

$$W = \left[\frac{100\,(t_h - t_m)}{(t_m - t_c)} - 100\right] cal/°C$$

$$W = \left[\frac{100\,(t_h - t_m)}{(t_m - t_c)} - 100\right] \times 4.184\, J/°C \text{ or } J/K$$

(A temperature difference of 1°C is equivalent to a temperature difference of 1K since the unit size in each scale is the same.)

Since the heat capacity of a calorimeter is also equivalent to the mass of water having the same heat capacity as the calorimeter, it can also be expressed in grams (g) or kilogram (kg).

Q. 7. Why energy changes in chemical reactions are usually expressed in terms of enthalpy rather than internal energy?

Ans.: Enthalpy (H) measures heat changes at constant pressure while internal energy (U) measures heat changes at constant volume. It is much more convenient to study a reaction at constant atmospheric pressure rather than at constant volume. Moreover, enthalpy gives the total energy change; the sum of internal energy and the energy change due to volume change at constant pressure (H = U + PV).

Q. 8. How do you define enthalpy of reaction and standard enthalpy of reaction?

Ans.: Enthalpy of reaction (ΔH) is defined as the sum of enthalpies of products minus sum of the enthalpies of reactants. Enthalpy changes of reactions determined at 25°C and 1 atmosphere pressure (101.325 kPa) are denoted by ΔH⁰ and are known as standard enthalpy changes.

Q. 9. Define enthalpy of solution.

Ans.: Enthalpy of solution is defined as the enthalpy change involved when one mole of solute dissolves in a specified quantity of solvent at a given temperature and pressure. Enthalpy of solution also depends on the quantity of solvent used to dissolve the solute. Usually the relative quantities of solute and solvent are expressed as mole ratio (1: n). The quantity so defined is the integral enthalpy of solution.

Q. 10. How do you differentiate between integral enthalpy of solution, differential enthalpy of solution and partial molar enthalpy of solution?

Ans.: Enthalpy of solution defined in Q.9 is the integral enthalpy of solution and this is the quantity which is usually measured. The partial molar enthalpy of solution is also called the differential enthalpy of solution. It is the enthalpy change per mole of solute when an infinitesimal amount of substance is dissolved in a large amount of solution of known concentration at constant temperature and pressure so that the solution concentration does not change appreciably. It can also be defined as the partial derivative of the total enthalpy of solution with respect to the molar concentration of one component of solution at constant pressure, temperature and concentration of all other components. In should also be noted that in integral enthalpy of solution the solute is dissolved in pure solvent whereas in differential enthalpy of solution the solute is dissolved in a solution of known concentration. Direct measurement of differential enthalpy of solution is not possible, it is usually determined indirectly from the knowledge of integral enthalpy of solution.

Q. 11. Why enthalpy of solution is positive for some substances while negative for others?

Ans.: Positive enthalpy change (ΔH) means that the process is endothermic; it proceeds with the absorption of heat. Negative enthalpy change, on the other hand, means that the process is exothermic and dissolution takes place with the evolution of heat. In order to understand this we have to think about various enthalpy changes that are involved in the process. The process of dissolution can be assumed to occur in two steps: breaking of crystal lattice and solvation of solute particles. The constituent particles in the crystal lattice may be ions in case of ionic solids and molecules in the case of molecular crystals. The breaking of the crystal lattice requires energy equal to the lattice energy of solute and is endothermic in nature. The other process, the solvation of solute particles is exothermic in nature. Thus the heat lost or gained during the dissolution process is roughly the sum of these two energy factors and may be written as

$\Delta H_{solution} = \Delta H_{lattice\ energy} + \Delta H_{solvation}$

The sign of $\Delta H_{lattice\ energy}$ is positive while that of $\Delta H_{solvation}$ is negative and therefore, the overall sign of enthalpy of solution depends on the relative magnitudes of these two terms on the right hand side. Generally lattice energy is larger than solvation energy and enthalpy of solution is positive but it is not always true.

Q. 12. Why enthalpy of solution depends on the concentration of solution? Define enthalpy of dilution.

Ans.: As discussed in Q. 11 above, the magnitude and sign of the enthalpy of solution is dependent on the relative magnitudes of lattice energy and solvation enthalpies. The amount of solvation energy released depends on the extent of solvation which further depends on the concentration of solution. Greater the dilution of a solution, greater is the concentration of solvent and greater is the extent of solvation and hence more energy is released. This is the reason why enthalpy of solution depends on the quantity of solvent. The enthalpy of dilution is defined as the change of enthalpy when a solution containing 1 mole of a solute is diluted from one concentration to another. The sign of enthalpy of dilution is negative in most cases.

Q. 13. Why dissolution of a salt in water results in the increase or decrease of the temperature of water?

Ans.: As already discussed in question 11, temperature will decrease on dissolution of salts with positive enthalpy of solution and increase on dissolution of salts with negative enthalpy of solution.

Q. 14. Define enthalpy of neutralization.

Ans.: The enthalpy of neutralization (ΔH_{neut}) is the change in enthalpy that occurs when one equivalent of an acid is neutralized by one equivalent of a base in dilute solution to form water and a salt. It is a special case of the enthalpy of reaction. The enthalpy of neutralization is always measured per mole of water formed since neutralization of one equivalent of acid by one equivalent of base produces one mole of water. Although the units of enthalpy of neutralization are often written as kJ/mole, the correct units are kJ/mole of water formed.

Q. 15. Why enthalpy of neutralization of an acid by a base is always negative?

Ans.: The neutralization reaction is basically the reaction between hydrogen and hydroxyl ions to form water. For example, when sodium hydroxide is neutralized by hydrochloric acid, the reaction can be written as $NaOH + HCl \rightarrow NaCl + H_2O$. Since NaOH, HCl and NaCl are all strong electrolytes, they exist in solution as the respective ions and therefore, the basic reaction is $OH^- + H^+ \rightarrow H_2O$. Since neutral water is more stable than hydrogen or hydroxyl ions, energy is released during the reaction. Hence the reaction is exothermic and enthalpy of neutralization is negative.

Q. 16. Why enthalpy of neutralization of all strong acids and strong bases are identical? What is this value?

Ans.: Since neutralization of all strong acids and strong bases involves the reaction between hydrogen and hydroxyl ions to form water, the enthalpy of neutralization of all strong acids and strong bases is identical. This value is reported to be -57.4 kJ mol^{-1}.

Q. 17. Ten kJ of heat is evolved when 100 mL 0f 0.2 N HCl is neutralized by 100 mL of 0.2N NaOH. What is the enthalpy of neutralization?

Ans.: Heat evolved = 10 kJ = 10×10^3 J
Number of g equivalents neutralized = $100 \times 0.2 \times 10^{-3}$
Enthalpy of neutralization = Heat evolved per g equivalent = $(10 \times 10^3)/(100 \times 0.2 \times 10^{-3}) = 5 \times 10^5$ J = 500 kJ/mol.

Q. 18. Why less heat is evolved in the neutralization of a weak acid by a strong base as compared to the neutralization of a strong acid?

Ans.: Neutralization of a strong acid takes place in a single step but the neutralization of a weak acid takes place in two steps. A weak acid has low degree of dissociation, it is only partially ionized and therefore, ionization and neutralization reactions take place side by side. For example,

(i) $CH_3COOH \rightleftharpoons CH_3COO^- + H^+$

(ii) $CH_3COO^- + H^+ + NaOH \rightarrow CH_3COONa + H_2O$

The first step (ionization) requires energy and is endothermic while the second step is exothermic. Since some energy is used up in the ionization of weak acid, less heat is evolved in the neutralization of a weak acid as compared to the neutralization of a strong acid.

Q. 19. What is enthalpy of ionization and what should be its sign. How enthalpy of ionization can be determined?

Ans.: The enthalpy of ionization is defined as the amount of heat required for ionization or dissociation of one mole of a compound into its constituent ions. The ionization reaction is always endothermic and the sign of enthalpy change is positive because energy is required to break a stable neutral molecule into ions which are less stable than the salt.

Enthalpy of ionization of a weak acid, for example can be determined from its enthalpy of neutralization (ΔH_{neut}). As already discussed in Q. 18, in the case of a weak acid such as acetic acid, ionization and neutralization reactions proceed side by side till the acid is completely neutralized. The experimentally determined enthalpy of neutralization of weak acid is equal to the sum of the enthalpy of ionization of weak acid and the enthalpy of neutralization of hydrogen and hydroxyl ions to form water which is known to be -57.4 kJ mol^{-1}. Therefore,

$$\Delta H_{neut} = \Delta H_{ionization} + \Delta H_{(H^+ + OH^- \rightarrow H_2O)}$$

$$\Delta H_{ionization} = \Delta H_{neut} - \Delta H_{(H^+ + OH^- \rightarrow H_2O)} = \Delta H_{neut} + 57.4 \; kJ \; mol^{-1}$$

In the weak acid-strong base neutralization reaction less heat is evolved; the sign of ΔH_{neut} is negative but its magnitude is less than 57.4 kJ mol^{-1} and therefore, $\Delta H_{ionization}$ has a positive value. The neutralization of a weak base by a strong acid can also be explained in the same way.

Q. 20. Explain the neutralization of a weak dibasic acid by a strong base.

Ans.: In the case of dibasic acid also ionization and neutralization reactions proceed side by side. However, both protons of the acid do not ionize in a single step; the ionization takes place in two steps. The two steps in the ionization of dibasic acid, oxalic acid are shown below. K_{a1} and K_{a2} are the dissociation constants for the two steps.

$H_2C_2O_4 \rightleftharpoons H^+ + HC_2O_4^-$, $K_{a1} = 6.5 \times 10^{-2}$, $HC_2O_4^- \rightleftharpoons H^+ + C_2O_4^{2-}$, $K_{a2} = 6.1 \times 10^{-5}$

It is seen that the second dissociation constant, K_{a2} is about three orders of magnitude lower than the first dissociation constant, K_{a1}. The basic difference is that the first proton is to be removed from a neutral molecule while the second proton is to be removed from a negatively charged ion. The removal of a positively charged proton from a negatively charged ion is much more difficult as compared to that from a neutral molecule. That is why both hydrogen ions are not available simultaneously and dissociation takes place in two steps. Neutralization of H^+ ions by reaction with base drives the ionization reaction. Thus ionization and neutralization reactions take place side by side.

Q. 21. Define Hess's law of constant heat summation.

Ans.: Hess's law of constant heat summation states that the total enthalpy change of a reaction is the same, regardless of whether the reaction is completed in one step or in several steps. In other words, enthalpy change is independent of the pathway between initial and final states and is thus a state function. It is another form of the first law of thermodynamics which is the law of conservation of energy.

Q. 22. How can we determine the enthalpy of hydration of an anhydrous salt?

Ans.: We can explain this by taking the example of copper sulphate. The enthalpy of solution of anhydrous copper (II) sulphate and copper(II) sulphate pentahydrate in water are determined. The enthalpy of hydration of copper(II) sulphate is evaluated by using Hess's Law of Constant heat (Enthalpy) Summation. The enthalpy of solution of anhydrous salt is equal to the sum of the enthalpy of hydration of anhydrous salt and the enthalpy of solution of the hydrated salt. So the enthalpy of hydration of anhydrous salt = enthalpy of solution of anhydrous salt - enthalpy of solution of hydrated salt. If the mole ratio is taken as 1:800 for the anhydrous salt, for hydrated salt it should be 1:795.
(i) $CuSO_4$ (Anhydrous) (s) + 800 H_2O (l) → $CuSO_4.5H_2O$ (aq) + 795 H_2O (l)
(ii) $CuSO_4.5H_2O$ (Hydrated) (s) + 795 H_2O (l) → $CuSO_4.5H_2O$ (aq) + 795 H_2O (l)
Subtracting, (i) – (ii) gives
$CuSO_4$ (Anhydrous) + 5 H_2O (l) → $CuSO_4.5H_2O$ (Hydrated)
Therefore, $\Delta H_{Hydration} = \Delta H_{Anhydrous} - \Delta H_{Hydrated}$

Q. 23. Define enthalpy of combustion.

Ans.: Enthalpy of combustion is the enthalpy change accompanying complete combustion of one mole of a substance. Combustion takes place with the evolution of heat and enthalpy of combustion is always negative.

Q. 24. Define enthalpy of formation of a compound.

Ans.: Enthalpy of formation of a compound is defined as the change in enthalpy when one mole of a compound is formed from its elements in their standard stable states. For example, the enthalpy of formation of carbon dioxide is the enthalpy change in the formation of 1 mole of gaseous carbon dioxide by reaction of solid carbon and gaseous oxygen. Enthalpy change is usually measured under standard conditions and is called the standard enthalpy of formation (ΔH^0_f). The standard conditions are a pressure of 1 atmosphere for gases and a concentration of 1 mol/L for all species in solution. Also the reactants must be pure substances in their most stable states under these conditions at a specified temperature. Most thermochemical data is tabulated at 25°C (298 K).

Q. 25. Can we measure standard absolute enthalpy of a compound?

Ans.: By convention enthalpies of formation of all elements in their standard stable states are taken as zero. Standard enthalpy of formation of a compound, ΔH^0_f can be written as

$$\Delta H^0_f = \sum H^0_f(Products) - \sum H^0_f(Reactants) = H^0_f(Compound)$$
$$- \sum H^0_f (Elements\ from\ which\ the\ compound\ is\ formed)$$

Since by definition, enthalpy of formation is the change in enthalpy when the compound is formed from its elements in their standard stable states and the standard enthalpies of formation of the elements is taken as zero by convention, the second term on the right hand side is zero and

$$\Delta H^0_f = \sum H^0_f(Products) = H^0_f(Compound).$$

Thus the standard enthalpy of formation of a compound is equal to the standard absolute enthalpy of the compound.

Q. 26. What is the significance of the sign of absolute enthalpy of a compound?

Ans.: Standard enthalpy of formation (ΔH^0_f), which is equal to the absolute enthalpy of the compound (H^0_f), may be positive or negative. Positive values indicate that the compound formed is less stable than the elements from which it is formed. Such compounds are called endothermic compounds. Negative values indicate that the compound formed is more stable than the elements from which it is formed. Such compounds are called exothermic compounds. For example, absolute enthalpies of $CO_2(g)$ and $NO_2(g)$ are $-$ 394kJ/mol and + 33.1 kJ/mol, respectively. Thus carbon dioxide is more stable and nitrogen dioxide is less stable than the elements from which they are formed.

Q. 27. Enthalpy changes in chemical reactions are temperature-dependent? Which equation describes this temperature-dependence?

Ans.: The extent of a chemical reaction that is the amount of products formed depends on temperature and since the enthalpy of products is different from the enthalpy of reactants, enthalpy of reaction is temperature-dependent. The effect of temperature on the enthalpy of reaction is given by Kirchoff's equation according to which

$$\frac{\Delta H_2 - \Delta H_1}{T_2 - T_1} = \Sigma C_p(Products) - \Sigma C_p(Reactants) = \Delta C_p$$

ΔH_2 and ΔH_1 are the enthalpies of reaction at temperatures T_2 and T_1, respectively and ΔC_p is the difference in the sum of the heat capacities of products and sum of the heat capacities of reactants. If enthalpy at one temperature is known that at another temperature can be calculated using Kirchoff's equation.

Q. 28. What are thermoanalytical techniques?

Ans.: Thermoanalytical techniques are a group of techniques where physical and chemical properties of materials are studied as a function of temperature or time when the material is subjected to uniform heating. Commonly used techniques include thermogravimetric analysis (TGA), differential scanning calorimetry (DSC) and differential thermal analysis (DTA). In thermogravimetric analysis, the substance is subjected to uniform heating and the change in the weight of substance is recorded as a function of temperature or time. In differential scanning calorimetry (DSC) the difference in the amount of heat required to increase the temperature of sample and reference is measured as a function of temperature. The basic principle involved is that a physical transformation, chemical reaction or conformational alteration in the sample will affect the heat required to raise its temperature to the same level as that of the reference. Differential thermal analysis (DTA) is similar to differential scanning calorimetry (DSC), the only difference is that rather than the heat flow difference, the temperature difference between the sample and reference is measured during constant heating or cooling rate or cycle.

Q. 29. What are thermometric titrations?

Ans.: Thermometric titrations involve the measurement of change in temperature of a solution during titration as a function of the volume of titrant added. Since all chemical reactions are associated with enthalpy changes resulting in increase or decrease in temperature of the reaction mixture, thermometric endpoint sensing can be applied to a wide range of titration types such as acid-base, redox, complexometric and precipitation titrations.

Q. 30. What is enthalpy of precipitation? Give an example.

Ans.: Enthalpy of precipitation is the quantity of heat evolved in the precipitation of one mole of a sparingly soluble substance on mixing dilute solutions of suitable electrolytes. For example on mixing 250 mL of 0.4 M $BaCl_2$ with 250 mL of 0.4 M Na_2SO_4, (0.4 x 250)/1000 = 0.1 mole of $BaSO_4$ will be precipitated. If H calories of heat is evolved, the enthalpy of precipitation of barium sulphate will be H/0.1 =10H calories = 10H x 4.184 Joule.

Q. 31. How can heat of solution of a substance be determined by solubility method?

Ans.: The variation of solubility, S of a substance with temperature is given by Van't Hoff equation.

$$\frac{d \ln S}{dT} = \frac{\Delta H}{RT^2}$$

where ΔH is the heat of solution of the substance, R is the molar gas constant and T is the absolute temperature. Assuming that ΔH is constant in the temperature range T_1 to T_2, integration of the above equation gives

$$\log \frac{S_2}{S_1} = \frac{\Delta H}{2.303 R} [\frac{T_2 - T_1}{T_1 T_2}]$$

S_2 and S_1 are the solubilities at temperatures T_2 and T_1, respectively. Experimental determination of S_2 and S_1 values enables calculation of enthalpy of solution, ΔH.

Q. 32. What is the primary application of Born-Haber cycle?

Ans.: The Born-Haber cycle, which is based on the Hess's law, is primarily used for the calculation of lattice energy of an ionic solid which cannot otherwise be measured directly. The energy required to break one mole of an ionic solid into gaseous ions or the energy released when gaseous ions combine to form one mole of an ionic solid is called lattice energy. Hess's law states that the standard enthalpy change of reaction is the algebraic sum of the standard enthalpies of reactions into which the overall reaction may be divided. The Born-Haber cycle applies to ionic solids only such as alkali halides. Let us take the example of sodium chloride. We start with elements in the natural stable state. The various reaction energies involved are given below.

(a) Na (s) + $\Delta H_{Sublimation}$ → Na (g)
(b) Na (g) + $\Delta H_{Ionization}$ → Na$^+$ (g)
(c) ½ Cl$_2$ + $\Delta H_{Bond\ Energy}$ → Cl (g)
(d) Cl (g) – Electron Affinity → Cl$'$ (g)
(e) Na$^+$ (g) + Cl$'$ (g) – Lattice Energy → NaCl (s)

The standard enthalpy of formation of sodium chloride, ΔH^0_f is the sum of the energies involved in reactions (a) to (e).

$$\Delta H^0_f = \Delta H_{Sublimation} + \Delta H_{Ionization} + \Delta H_{Bond\ Energy} + Electron\ Affinity + Lattice\ Energy$$

The first three enthalpy changes are positive since energy is required for these processes while electron affinity and lattice energy are negative, the energy is released in these two steps. Lattice energy is calculated by subtracting the first four energies from the enthalpy of formation of compound. In the above equation all energies except lattice energy can be experimentally determined or found from the literature. For sodium chloride the lattice energy is so high that more energy is released in this step than is required in all preceding steps. Lattice energy is a measure of the strength of bonds in an ionic compound.

8. Acid-base indicators

Q. 1. What is the role of an indicator in a titration?

Ans.: An indicator is a substance which gives visual indication about the completion of the reaction. The end point of a titration can be determined easily with the help of an indicator.

Q. 2. What types of indicators are generally used?

Ans.: Generally used indicators can be divided into five groups: acid-base, oxidation-reduction, complexometric, adsorption and chemiluminescent indicators.

Q. 3. What are acid-base indicators?

Ans.: Usually acid-base indicators are weak organic acids or bases which change color with change in the pH of solution. They have the distinct property that the ionized and unionized forms have different colors.

Q. 4. Explain the working of an indicator which is a weak acid.

Ans.: Consider a weak acid indicator, HIn. In aqueous medium the following equilibrium is established.

$$HIn\ (aq) + H_2O\ (l) \rightleftharpoons In^-\ (aq) + H_3O^+\ (aq)$$

Let us say that the unionized acidic form, HIn has color A while the basic ionized form has color B. At low pH, the concentration of H_3O^+ is high, the equilibrium lies to the left and the solution has color A. At high pH, the concentration of H_3O^+ is low, the equilibrium lies to the right and the solution has color B. For example, the unionized form of phenolphthalein is colorless in acidic medium while the ionized form is pink (magenta) in basic medium.

Q. 5. Explain the working of an indicator which is a weak base.

Ans.: Consider a weak base indicator, BOH which ionizes in aqueous medium as follows.

BOH (aq) ⇌ B⁺ (aq) + OH⁻ (aq)

Let us say that the unionized form, BOH has color A and the ionized form, B⁺ has color B. At high pH (basic medium), excess of hydroxyl ions shift the equilibrium to the left and the solution has color A. At low pH, hydroxyl ions produced are neutralized by hydrogen ions from acid, the equilibrium lies to the right and the solution has color B. For example, the unionized form of methyl orange is yellow in basic medium while the ionized form is red in acidic medium.

Q. 6. What is the Ostwald theory of acid-base indicators?

Ans.: According to Ostwald theory, the colour change of an acid-base indicator is due to its ionization since ionized and unionized forms have different colours. Acid-base indicators are weak organic acids or bases. An acidic indicator has low ionization in acidic medium and high ionization in alkaline medium while a basic indicator has high ionization in acidic medium and low ionization in alkaline medium. The colour change is due to the shift in ionization equilibrium to the right or left in acidic or basic medium. The working of a weak acid or weak base indicator, explained in Q. 4 and Q. 5, is in accordance with the Ostwald theory of acid-base indicators.

Q. 7. What is the quinonoid theory of acid-base indicators?

Ans.: According to the quinonoid theory, acid-base indicators exist in two tautomeric forms having different structures. One form mainly exists in acidic medium and the other in alkaline medium. The two forms have different colors and are in equilibrium due to the inter conversion of one tautomeric form into other. During titration, the change in pH converts one tautomeric form into other and thus the color change occurs.

Q. 8. What criteria helps us decide which indicator should be used for a given acid-base titration?

Ans.: An acid-base indicator is suitable for a particular titration if it changes color near the equivalence point of the titration. Equivalence point, also called end point of titration is the point where acid and base are mixed in exactly equal proportions. The indicator should change color as close as possible to the equivalence point. For example phenolphthalein, a commonly used indicator is a weak acid; the unionized form of which is colorless whereas the ionized form is bright pink. It changes color in the pH range 8.3 – 10 and is suitable for a titration involving strong base such as a strong acid-strong base and weak acid-strong base titration. However if the base is weak, the pH prevailing near the end point is much lower and phenolphthalein is not a suitable indicator since the hydroxyl ions furnished by the weak base at the end point are not enough to shift the equilibrium sufficiently towards the right to raise the pH to 8.3. Methyl orange which is a weak base and changes color in the pH range 3.1 - 4.4 is a suitable indicator for titrations in which the base is weak such as a strong acid-weak base titration.

Q. 9. Which indicator is suitable for a weak acid-weak base titration?

Ans.: When a weak acid is titrated with a weak base, the pH change is too gradual close to the end point. No indicator is suitable for this titration since the color change of the indicator near the end point is not sharp. That is why it is generally recommended that a weak acid should not be volumetrically titrated with a weak base.

Q. 10. Why strong acids and strong bases are generally not suitable as acid-base indicators?

Ans.: In the case of a weak acid or a weak base indicator, the ionized and unionized forms exist in equilibrium. When such an indicator is added to an acid or base, shift in equilibrium may occur resulting in color change since ionized and unionized forms of indicator have different colors. Strong acids and strong bases, on the other hand remain practically fully ionized at all pH values of interest, the ionized and unionized forms do not exist in equilibrium and therefore, they are not suitable as acid-base indicators.

Q. 11. What do you understand by the K_{In} and pK_{In} values of indicators?

Ans.: K_{In} is the indicator dissociation constant just like the acid dissociation constant and $pK_{In} = -\log K_{In}$. In the case of a weak acid indicator HIn for example, the law of chemical equilibrium can be applied to the following equilibrium

HIn (aq) + H_2O (l) ⇌ In^- (aq) + H_3O^+ (aq)

$$K_{In} = \frac{[In^-][H_3O^+]}{[HIn]} \text{ and } -\log K_{In} = -\log[H_3O^+] - \log\frac{[In^-]}{[HIn]}$$

$$pK_{In} = pH - \log\frac{[In^-]}{[HIn]} \text{ or } pH = pK_{In} + \log\frac{[In^-]}{[HIn]} = pK_{In} + \log\frac{[Ionized]}{[Unionized]}$$

The above equation is called Henderson-Hasselbatch equation.

When $[In^-] = [HIn]$, $pK_{In} = pH$. Since ionized and unionized species, $[In^-]$ and $[HIn]$ have different colors, pK_{In} is equal to the pH of the solution at its turning point. It is the pH at which half of the indicator is in its acid form and the other half is in its conjugate base form.

Q. 12. What is the significance of the pk_{In} value of an indicator?

Ans.: According to the Henderson-Hasselbatch equation, pK_{In} value is equal to the pH when concentrations of ionized and unionized forms of the substance are equal. Since the acid and conjugate base forms, that is, the ionized and unionized forms of the indicator have different colors, the colour change will occur at a pH (= pK_{In}) where 50% of both forms are present.

Q. 13. What do you understand by pH indicator range?

Ans.: As explained above the color change occurs at the turning point. If pH is above the pK_{In} value, the concentration of the conjugate base is greater than the concentration of the acid, and the color associated with the conjugate base dominates. If pH is below the pK_{In} value, the color of the acid from dominates. Usually, the color change is not instantaneous at the pK_{In} value but a pH range exists where a mixture of colors is present. This pH range varies for different indicators. An indicator is most effective if the color change is distinct and over a small pH range. For most indicators the approximate range is within ± 1units of the pK_{In} value. This assumes that solutions retain their color as long as at least 10% of the other species persists. For example, if the concentration of the conjugate base is 10 times greater than the concentration of the acid, their ratio is 10:1, and consequently the pH is $pK_{In} + 1$. Conversely, if a 10-fold excess of the acid occurs with respect to the base, the ratio is 1:10 and the pH is $pK_{In} - 1$.

Q. 14. What is the pH range and extreme colors in acidic and alkaline solutions for some commonly used indicators?

Ans.: Table 7 lists some commonly used acid-base indicators along with their pH ranges and the corresponding color change in acidic and basic solutions.

Q. 15. What is litmus paper and how it is used to test the acidity or alkalinity of a solution.

Ans.: Litmus paper is a filter paper which has been treated with a mixture of natural water-soluble dyes obtained from lichens. Blue litmus paper turns red in acidic medium and red litmus paper turns blue in basic or alkaline medium. The color change occurs over the pH range 4.5–8.3. Neutral litmus paper is purple. Litmus can also be prepared as an aqueous solution that functions in the same way.

Q. 16. Why only a small amount of indicator should be used in acid-base titrations?

Ans.: Only a small amount of indicator should be used in acid-base titrations. This is because indicators are also weak acids or bases and therefore, larger amount will cause error in the estimation of acidity or basicity.

Q. 17. How common acid-base indicators, phenolphthalein and methyl orange are prepared and what concentrations are generally used?

Ans.: Phenolphthalein has very low solubility in water. It is sufficiently soluble in ethanol, a water-miscible solvent and therefore, phenolphthalein indicator solution is generally prepared in 50% aqueous ethanol. Methyl orange, however, is sufficiently soluble in water and methyl orange indicator solution is prepared in water. Concentration used is usually 0.5 or 1% for phenolphthalein and 0.1% for methyl orange.

Table 7: pK_{In}, pH range and the color change for some commonly used acid-base indicators.

Indicator	pH range	Color	
		Acidic solution	Basic solution
Cresol Red (acid)	1.2 – 1.8	Red	Yellow
Thymol Blue (acid)	1.2 – 2.8	Red	Yellow
Methyl Orange	3.1 - 4.5	Red	Yellow
Methyl Red	4.2 - 6.3	Yellow	Red
Bromothymol Blue	6.0 - 7.6	Yellow	Blue
Phenol Red	6.4 - 8.2	Yellow	Red
Cresol Red (Base)	7.0 – 8.1	Yellow	Red
Thymol Blue (Base)	8.1 - 9.6	Yellow	Blue
Phenolphthalein	8.0 – 9.8	Colorless	Pink

* The data reported in the above Table has been taken from Principles of Physical Chemistry by B. R. Puri, L. R. Sharma and M. S. Pathania, p 650. Vishal Publishing company, Delhi, 43rd Edition.

Q. 18. What is a universal indicator?

Ans.: A universal indicator is a mixture of indicators which give a gradual change in color over a wide pH range. The pH of a solution can be approximately identified when a few drops of universal indicator are mixed with the solution.

Q. 19. What is the difference between acid-base indicators and oxidation-reduction indicators?

Ans.: In an acid-base indicator the ionized and unionized forms of the substance have different color while in an oxidation-reduction indicator, also called redox indicator, the oxidized and reduced forms of the compound have different colors. A redox indicator changes color at a specific electrode potential and the redox reaction is reversible. Methylene blue, diphenylamine, ferroin and starch are some of the examples of redox indicators.

Q. 20. What are the everyday uses of indicators?

Ans.: Some of the everyday uses of indicators include testing the acidity and basicity of soils, checking the pH of swimming pool water, monitoring the acidity or alkalinity of liquid wastes to be discharged into the sewerage system. For everyday use, pH papers available in different pH-ranges are also quite useful.

9. pH metry

Q. 1. What is the correct way to define pH?

Ans.: pH is a logarithmic scale used to specify the acidity or basicity of an aqueous solution. It is defined as

$$pH = -\log a_{H^+}$$

a_{H^+} is the activity of hydrogen ions in solution. This is the precise way to define pH value. However, many times approximate definition is also used according to which $pH = -\log [H^+]$, where $[H^+]$ is the molar concentration of hydrogen ions in solution. In aqueous solution the free cation of acid is mostly present as hydrated proton, H_3O^+ but for convenience it is often referred to as hydrogen ion, H^+.

Q. 2. What is pH scale and what does it represent?

Ans.: pH is a numeric scale used to specify the acidity or basicity of an aqueous solution. The pH scale usually ranges from 0 to 14. This is because the concentration of hydrogen ions in most aqueous solutions lies between 1 and 10^{-14} M. At a pH of zero, the concentration of hydrogen ions is 1M while at a pH of 14 the concentration of hydroxyl ions is 1M. However, one must keep in mind that at concentration as high as 1M activity is not equal to concentration and the above statement is only approximately true. A pH value of 7 indicates that the concentration of hydrogen ions is equal to the concentration of hydroxyl ions and the solution is neutral. In an acidic solution, the concentration of hydrogen ions is more than the concentration of hydroxyl ions and the pH is less than 7 while in a basic solution, the concentration of hydroxyl ions is more than the concentration of hydrogen ions and the pH is more than 7.

Q. 3. Can pH be zero, less than zero (negative) or greater than 14?

Ans.: Yes, pH value can be zero, less than zero (negative) or greater than 14. pH should be zero if the hydrogen ion concentration is 1M; less than zero (negative) if the hydrogen ion concentration is more than 1M and greater than 14 if the hydroxyl ion concentration is more than 1M. For example, pH values of 1 M and 10 M HCl solutions should be approximately zero and minus 1, respectively while that of 10 M NaOH solution should be approximately 15, assuming complete dissociation. However, one must remember that pH can only be precisely calculated using activities rather than concentrations and only at low concentrations activity coefficients are close to unity and activities are close to concentrations. At high concentrations the activity coefficients are less than zero and activities are less than concentrations and therefore, at these concentrations activities rather than concentrations should be used in the calculation of pH value.

Q. 4. How do you define the ionic product of water and what is its significance?

Ans.: The ionic product of water (K_w) is defined as the product of the concentration of hydrogen and hydroxyl ions in an aqueous solution, K_w = [H$^+$][OH$^-$] = 10^{-14} at 25°C. Again the precise way to write this relation is K_w = [H$_3$O$^+$][OH$^-$] since hydrogen ion exists as hydronium ion in aqueous solution. The significance of the ionic product of water is that in an aqueous solution, the concentration of hydrogen and hydroxyl ions are related in such a way that if one increases, the other decreases and vice versa so that the product remains constant.

Q. 5. Why and how the ionic product of water varies with temperature?

Ans.: Water is a weak electrolyte and exists in equilibrium with hydrogen and hydroxyl ions. Auto-ionization of water can be written as H$_2$O $\rightleftharpoons$ [H$^+$] + [OH$^-$] or more precisely as 2H$_2$O $\rightleftharpoons$ [H$_3$O$^+$] + [OH$^-$]. Thus concentration of hydrogen and hydroxyl ions depends on the extent of ionization of water which increases with increase in temperature. Therefore, the ionic product of water, which is the product of hydrogen and hydroxyl ion concentrations, also increases with increase in temperature. On increasing the temperature from 0 to 60°C, K_w increases from approximately 10^{-13} to 10^{-15}.

Q. 6. The pH of pure water is always 7? Explain.

Ans.: No, pH of pure water is not exactly 7 always, it is only approximately 7. Since the concentration of hydrogen ions available from water varies with temperature, pH also varies with temperature. Increase in the temperature of water from 0 to 100°C increases the ionic product of water from 0.114×10^{-14} to 0.513×10^{-12} and the pH of pure water decreases from 7.47 to 6.14. However, pure water is always neutral irrespective of temperature since water always produces equivalent concentrations of hydrogen and hydroxyl ions.

Q. 7. What is the pH of water at physiological temperature?

Ans.: Physiological temperature is the body temperature which is 37°C. We know that the ionization of water and hence the ionic product of water, K_w varies with temperature. At 37°C, $K_w = [H^+][OH^-] = 2.4 \times 10^{-14}$. Since water produces equivalent amounts of hydrogen and hydroxyl ions, $[H^+] = (2.4 \times 10^{-14})^{1/2} = 1.55 \times 10^{-7}$ and $pH = -\log(1.55 \times 10^{-7}) = 6.81$.

Q. 8. Are there any hydroxyl ions in an acidic solution? Are there any hydrogen ions in an alkaline solution?

Ans.: Yes acidic solutions do contain hydroxyl ions and alkaline solutions contain hydrogen ions. This is because neutral water contains both hydrogen and hydroxyl ions at a concentration of 10^{-7} M and the product of hydrogen and hydroxyl ion concentration is a constant equal to 10^{-14} at 25°C. So the concentration of hydrogen and hydroxyl ions are related in such a way that if one increases, the other decreases and vice versa so that the product remains constant. If the concentration of hydrogen ions is more than that of hydroxyl ions, the solution is acidic and if the concentration of hydroxyl ions is more than that of hydrogen ions, the solution is basic. For example, in 0.1M HCl, the concentration of hydrogen ions is 10^{-1} M (assuming complete dissociation) and that of hydroxyl ions is 10^{-13}M. Similarly, the concentration of hydroxyl ions is 10^{-1}M in 0.1M NaOH solution and that of hydrogen ions is 10^{-13}M (assuming complete dissociation).

Q. 9. What is the concentration of hydrogen and hydroxyl ions in a solution of pH 5 at 25°C?

Ans.: Ionic product of water, $[H^+][OH^-] = 10^{-14}$ M at 25°C. In a solution of pH 5, $[H^+] = 10^{-5}$M, assuming complete dissociation. Since $[H^+][OH^-] = 10^{-14}$, $[OH^-] = 10^{-9}$ M.

Q. 10. What is the concentration of hydrogen and hydroxyl ions in a solution of pH 10 at 25°C?

Ans.: Since pH = $-\log [H^+]$, in a solution of pH 10, $[H^+] = 10^{-10}$ M. Since $[H^+][OH^-] = 10^{-14}$, $[OH^-] = 10^{-4}$ M.

Q. 11. Define pOH and the relation between pH and pOH of a solution. The pOH value of a solution is 10. What is its pH value?

Ans.: Just like pH, pOH can be defined as pOH = $-\log[OH^-]$ or more precisely pOH = $-\log a_{OH^-}$ where a_{OH^-} is the activity of hydroxyl ions in solution. We know that the ionic product of water, $[H^+][OH^-] = 10^{-14}$ at 25°C. Take log on both sides and then put a negative sign. $-\log [H^+] + (-\log[OH^-]) = 14$ or pH + pOH = 14. Thus if pOH is 10, pH of solution = 14 − 10 = 4.

Q. 12. What will be the pH of 0.1N HCl solution diluted 10 times with water?

Ans.: Ten times dilution of 0.1N HCl produces 0.01N or 10^{-2} N HCl. The concentration of hydrogen ions, assuming complete dissociation is 10^{-2} N and the pH of the acid solution is 2.

Q. 13. What will be the pH of 0.1N HCl solution diluted 10^8 times with water?

Ans.: By applying a dilution factor of 10^8, the concentration of diluted HCl solution would be $0.1/10^8 = 10^{-9}$ and the calculated pH of this solution is 9. However, this cannot be true since an acid solution cannot have a pH greater than 7. We know that even in neutral water, the concentration of hydrogen ions is 10^{-7} which is 100 times more. Therefore, in this case concentration of H^+ from water dissociation cannot be neglected. Therefore, the pH of this solution will be close to 7.

Q. 14. A pH meter is basically a potentiometer? Is it true or false?

Ans.: This is true. A pH meter is basically a potentiometer. The electrode assembly is just like a galvanic cell and the working principle is same as that of a potentiometer. The only difference is that there is an internal calibration in the instrument which enables potential difference to be converted to pH value.

Q. 15. What is the working principle of a pH meter?

Ans.: Since a pH meter is basically a potentiometer, the parameter which is measured experimentally is potential difference. It measures the potential difference between a glass electrode which is reversible w.r.t hydrogen ions and a reference electrode such as calomel electrode or platinum electrode. Just like a potentiometer, it works on the poggendorf's compensation principle according to which the unknown potential is opposed to another variable known potential till the two are balanced and no current flows through the galvanometer. There is an internal calibration in the pH meter which converts the experimentally determined potential difference to pH using Nernst equation.

Q. 16. What is the construction of a glass electrode?

Ans.: A glass electrode is a type of ion-selective electrode that consists typically of a glass tube, sealed at the bottom, with a doped glass membrane bulb containing a solution of constant pH (usually a chloride buffer of pH 7) and a silver-silver chloride or platinum reference electrode. The construction of the electrode may vary from one manufacturer to another. This is immersed in the unknown solution, usually along with a calomel electrode, for determining the pH of this solution. The doped glass membrane is sensitive to hydrogen ions. Due to the difference of hydrogen ion concentration inside and outside the glass bulb, a potential is generated at the electrode which can be measured. The majority of pH electrodes available commercially are combination electrodes that have both glass electrode which is H^+ ion sensitive electrode and an additional reference electrode placed together in a single tube. This is done only for convenience of use.

Q. 17. How the glass electrode should be cleaned and stored?

Ans.: The pH electrode should be cleaned periodically to prevent buildup of material on the glass surface. A coating on the glass bulb will cause errors in measurement. Cleaning solution will depend on the type of coating. In general, dipping in 0.4 M HCl, warm 4 M KCl solution, warm pH 4 buffer or warm detergent solution for half an hour followed by thorough rinsing with water serves the purpose. However, it should be kept in mind that while cleaning only the glass bulb should be immersed in the cleaning solution. For best results it is always advisable to keep the glass electrode bulb hydrated. The electrode should always be cleaned and placed in the storage solution. Generally recommended storage solutions are 3 or 4 M KCl, pH 4 or 7 buffer solution. Most electrodes are provided with a protective cap which can be filled with the storage solution for long time storage. The hole in the electrode which provides junction between inner solution and the test solution should also be closed when the electrode is not in use. Do not store the electrode in distilled or deionized water because it will cause ions to leach out of the glass bulb thereby affecting the electrode performance.

Q. 18. Why pH meter is to be standardized or calibrated before use?

Ans.: The pH meter has a membrane that allows H⁺ ions to pass though. The flow of current generates voltage which is measured by the meter. The developed voltage is not very precise. It varies from electrode to electrode because electrodes of different makes do not have identical characteristics. It also changes with aging of the electrode and coating of membrane with use. It can also vary with other specifications of the instrument. This difference between the theoretical and actual behavior of a pH electrode must be compensated. A calibration is therefore, required to obtain correct pH value of the test solution.

Q. 19. What is the procedure used for calibration of a pH meter?

Ans.: A pH meter requires calibration to produce accurate pH readings. The calibration is done by immersing its electrode(s) into standard buffers and by adjusting the meter accordingly. One (1-) or two (2-) point calibration can be done. In 1-point calibration, a buffer with a pH close to the expected sample pH is used. A 2-point calibration is usually recommended for accurate data. The meter is first calibrated with a standard buffer of pH 7. The calibration is then checked by measuring the pH of a second buffer. The measured pH of the second buffer should be within ± 1 pH units of the actual value. The second buffer selected for this purpose should have a pH as close as possible to the suspected pH of the sample. Usually if the test solution is acidic, a buffer of pH 4 is used while if it is alkaline, a buffer of pH 9.2 or 10 can be used. Buffer tablets of pH 4, 7 and 9.2 are also available. Since pH measurements are affected by temperature, the temperature must remain constant during the time of the calibration. Modern pH meters have built-in thermometers and automatically correct their own pH measurements as the temperature changes.

Q. 20. Define the terms pK_a and pK_b. What does K stand for in these symbols?

Ans.: pK_a and pK_b values usually refer to weak acids and weak bases, respectively. K in these symbols stands for the dissociation constant of the acid or base. $pK_a = -\log K_a$ and $pK_b = -\log K_b$.

Q. 21. pK_a values for acids A and B are 4 and 6, respectively. Which acid is stronger and why?

Ans.: By definition, $pK_a = -\log K_a$. For acid A, $pK_a = 4$, $K_a = 10^{-4}$. Similarly for acid B, $pK_a = 6$, $K_a = 10^{-6}$. It is apparent that smaller pK_a value corresponds to higher dissociation constant and vice versa. Also stronger the acid higher is the dissociation constant. Therefore, acid A with smaller pK_a value and higher dissociation constant is stronger than acid B.

Q. 22. pK_b values for bases A and B are 5 and 7, respectively. Which is the weaker base and why?

Ans.: By definition, $pK_b = -\log K_b$. For base A, $pK_b = 5$, $K_b = 10^{-5}$. Similarly for base B, $pK_b = 7$, $K_b = 10^{-7}$. It is apparent that smaller pK_b value corresponds to higher dissociation constant and vice versa. Also stronger the base, higher is the dissociation constant. Therefore, base B with a higher pK_b value and lower dissociation constant is weaker than base A with a lower pK_b value.

Q. 23. What is the relation between pK_a and pK_b values for a given weak acid? What are pK_b values for acids A and B in Q. 21?

Ans.: For a given weak acid, K_a and K_b are the dissociation constants of the weak acid and its conjugate base, respectively. For example, a weak acid HA dissociates in water to hydronium ion, [H_3O^+] and its conjugate base A^-. The conjugate base, A^- reacts with water to form acid HA and hydroxyl ion. Therefore, neglecting the concentration of water which is present in excess K_a and K_b can be written as $K_a = [H_3O^+][A^-]/[HA]$ and $K_b = [HA][OH^-]/[A^-]$. The product $K_a K_b = [H_3O^+][OH^-] = K_w$, the ionic product of water. If we take log on both sides and put a negative sign, we get $pK_a + pK_b = pK_w = 14$ since $pK_w = -\log K_w$ and $K_w = 10^{-14}$ at $25°C$. For acids A and B in Q. 21 with pK_a values 4 and 6, pK_b values are 10 and 8, respectively.

Q. 24. Write Hendersen's -Hasselbalch equation.

Ans.: Hendersen's-Hasselbalch equation is obtained by rearrangement of the expression for the dissociation constant of a weak acid or weak base. For a weak acid HA for example, $K_a = [H^+][A^-]/[HA]$. If we take log on both sides and put a negative sign, we get pH = pK_a + log $[A^-]/[HA]$ or pH = pK_a + log [Salt form]/[Acid form]. Similarly for a weak base BOH, $K_b = [B^+][OH^-]/[BOH]$ and pOH = pK_b + log $[B^+]/[BOH]$ or pOH = pK_b + log [Salt form]/[Base form].

Q. 25. How is Hendersen's Hasselbalch equation useful?

Ans.: Hendersen's-Hasselbalch equation can be used to calculate pH of a buffer solution formed by mixing known amounts of a weak acid and its salt or a weak base and its salt. It can also be used to determine the pK_a and pK_b values of the weak acid or base used. During neutralization reaction when base is added to acid, salt and water are formed. At complete neutralization all the acid present is converted to salt. At half neutralization, half of the acid is converted to salt, [Salt] = [Acid] and pH = pK_a. Similarly when acid is added to base, at half neutralization of the base, [Salt] = [Base] and pOH = pK_b.

Q. 26. What is a pH titration curve or pH neutralization curve and how it is determined?

Ans.: In pH titration or pH neutralization curve the pH of the reaction mixture is used to monitor the acid-base neutralization reaction. If the pH of an acid solution is plotted against the amount of base added during a titration, the shape of the graph is called pH titration curve. All the acid-base titration curves follow the same basic shape. At the beginning the pH rises slowly on the addition of strong base but as the solution nears the point where all the hydrogen ions are neutralized by base, the pH rises sharply and then levels out as the solution becomes more basic. The shape of pH neutralization curve has been explained in detail subsequently.

Q. 27. What do you understand by the equivalence point of an acid-base pH-neutralization curve?

Ans.: pH-neutralization curve is a graph of the pH as a function of the amount of titrant (acid or base) added. The equivalence point (end point) of the titration is the point at which the amount of titrant added in just enough to neutralize all acid/base present in the reaction mixture with no titrant left over. In other words, at the equivalence point, the number of moles of titrant added so far corresponds exactly to the number of moles of substance being titrated according to the reaction stoichiometry. It is found that the pH increases slowly at first, then rapidly as the titration nears the equivalence point. The point of inflexion corresponds to the equivalence point.

Q. 28. Why there is a rapid change in pH as the equivalence point is approached?

Ans.: When an acid is titrated with a base decrease in the concentration of hydrogen ions results in increase in pH. The pH increases slowly at first because the pH scale is logarithmic, which means that a pH of 1 will have 10 times the hydrogen ion concentration than a pH of 2. Thus, as the hydrogen ion is initially removed, it takes a lot of base to change its concentration by a factor of 10, but as more and more hydrogen ion is removed, lesser base is required to change its concentration by a factor of 10. Near the equivalence point, a change by a factor of 10 occurs very quickly, which is why the graph is extremely steep at this point. Thereafter the curve levels out because it will again take a lot of base to increase the hydroxide ion concentration by 10 fold to change the pH significantly. The equivalence point is exactly at the center of the most vertical portion of the plot, also called the inflexion point.

Q. 29. How pK_a value of a weak acid can be determined from weak acid-strong base pH-neutralization curve?

Ans.: According to Hendersen's equation, $pH = pK_a + \log [A^-]/[HA]$ or $pH = pK_a + \log$ [Salt form]/[Acid form]. Neutralization of an acid and a base always produces salt and water. At complete neutralization, whole of the acid has been converted to salt while at half neutralization half of acid has been converted to salt and the half is present as acid. Under these conditions, the second term in the Hendersen's equation becomes zero and pH = pK_a. The pH corresponding to half neutralization of acid in the pH-neutralization curve is equal to the pK_a value.

Q. 30. How pK_b value of a weak base can be determined from weak base-strong acid pH-neutralization curve?

Ans.: pK_b value of a weak base can be determined in the same way as described above for a weak acid. According to base form of Hendersen's equation, pOH = pK_b + log [B^+] / [BOH] or pOH = pK_b + log [Salt form]/ [Base form]. If a weak base is titrated with a strong acid, again at half neutralization of base in the pH-neutralization curve, pOH = pK_b. The pH value corresponding to half neutralization can be converted to pOH by the relation pH + pOH = 14 and hence pOH = pK_b = 14 - pH at half neutralization.

Q. 31. For precise determination of end point in a weak acid-strong base or strong acid-weak base pH-neutralization curve sometimes a derivative plot is drawn. What is a derivative plot and why it is useful?

Ans.: In the case of a strong acid-strong base pH-neutralization curve, the rise in pH near the equivalence point is quite sharp, the curve is almost vertical in this region and the equivalence point can easily be read from the curve. The corresponding pH rise in the case of a weak acid-strong base pH-neutralization curve is not very sharp and therefore, it is difficult to determine the equivalence point accurately from the curve. Same is true for a strong acid-weak base titration curve. A derivative plot helps in the precise determination of end point in such cases. First derivative plot represents the change in slope of the titration curve as a function of the added volume of base. It is obtained by plotting the change in pH divided by the change in volume, $(pH_2 - pH_1)/(V_2 - V_1) = \Delta pH/\Delta V$ versus the average volume $[(V_2 + V_1)/2]$ of titrant added. The plot shows steep rise followed by a steep decline, the volume of base corresponding to maximum $\Delta pH/\Delta V$ value (the highest point on the graph) represents the equivalence point of the titration. The precision of the end point can be further improved by drawing a second derivative plot which is a plot of the second derivative, $\Delta^2 pH/\Delta V^2$ against volume of titrant. The $\Delta^2 pH/\Delta V^2$ values can easily be obtained from the first derivative plot. The second derivative should be zero at the end point.

Q. 32. How can we calculate the pH of a) 0.01N HCl, b) 0.01N CH_3COOH solution?

Ans.: a) pH of 0.01N HCl: At this concentration, if HCl is assumed to be fully dissociated, the concentration of hydrogen ions is 10^{-2} and pH of 0.01N HCl = 2.
b) pH of 0.01N CH_3COOH: Acetic acid, being a weak acid, is only slightly ionized even at 0.01N concentration. To calculate the pH of acetic acid solution we need to know its degree of dissociation (α) at this concentration (c). The fraction of total number of molecules of electrolyte dissolved that ionizes at equilibrium is known as the degree of dissociation or degree of ionization. Concentration of hydrogen ions = αc and pH of solution = $- \log (\alpha c)$. The degree of dissociation α at this concentration of acid is close to 0.04 and pH = 3.4. Thus if we compare HCl and CH_3COOH solutions, weaker acid with lower degree of dissociation has a higher pH that is less acidic than the stronger acid at the same concentration.

Q. 33. How can we calculate the pH of a) 0.01N NaOH, b) 0.01N NH_4OH?

Ans.: a) pH of 0.01N NaOH: At a concentration of 0.01N NaOH, if we assume complete dissociation, $[OH^-] = 10^{-2}$. Ionic product of water, $[H^+][OH^-] = 10^{-14}$ M at 25°C. So $[H^+] = 10^{-14}/10^{-2} = 10^{-12}$ and pH = $- \log [H^+]$ = 12.
b) pH of 0.01N NH_4OH: Ammonium hydroxide, being a weak base, is only slightly ionized even at 0.01N concentration. To calculate the pH of ammonium hydroxide solution we need to know its degree of dissociation (α) at this concentration (c). Concentration of hydroxyl ions = αc. So $[H^+] = 10^{-14}/\alpha c$ and pH = $- \log [H^+]$. The degree of dissociation of ammonium

hydroxide (α) at this concentration is close to 0.04. $[H^+] = 2.5 \times 10^{-11}$ and pH = 10.6. Thus if we compare NaOH and NH_4OH solutions, weaker base with lower degree of dissociation has a lesser pH; it is more acidic that is less basic than the stronger base at the same concentration.

Q. 34. What volume of 0.02N HCl should be used for complete neutralization of 50 mL of 0.01N NH_4OH.

Ans.: Although ammonium hydroxide is a weak base, it is fully neutralized. Ionization reaction, $NH_4OH \rightleftharpoons NH_4^+ + OH^-$ and neutralization reaction, $NH_4^+ + OH^- + HCl \rightarrow NH_4Cl + H_2O$ take place side by side. Since HCl concentration is double the base concentration, half volume, i.e., 25 mL of HCl should be used.

Q. 35. Do all liquids have a pH value?

Ans.: No, all liquids do not have a pH value. pH is a measure of the hydrogen or hydronium ion concentration of a solution. So if a liquid does not contain ionizable proton it will not have any pH. For example organic liquids such as hexane, vegetable oil, liquid metals, molten salts etc. Usually pH has significance only in aqueous solutions.

10. Colorimetry

Q. 1. What is colorimetry?

Ans.: Colorimetry is a widely used technique in physical, analytical and biological chemistry to determine the concentration of a colored substance in solution. It is based on the absorption of light of a specific wavelength in the visible range of electromagnetic spectrum by a colored sample.

Q. 2. Why different solutions show different colors?

Ans.: The visible light spectrum consists of a range of frequencies, each of which corresponds to a specific color. The color of a solution is a direct result of the frequency of light which is not absorbed by the sample. So, if a solution absorbs all of the frequencies of visible light except for the frequency associated with blue light, then the solution will appear blue and so on.

Q. 3. Which instrument is used in colorimetry?

Ans.: The devise used in colorimetry is called a colorimeter. Colorimeter measures the absorbance (optical density) and/or transmittance of light by the colored solution at a specified wavelength.

Q. 4. What are the essential parts of a colorimeter?

Ans.: The essential parts of a colorimeter are i) a light source (usually a low voltage filament lamp), ii) an aperture (aperture is a hole or an opening through which light travels), iii) a set of colored filters (Optical filters are devices that selectively transmit light of different wavelengths), iv) a cuvette containing the sample solution, v) a photodetector (usually a light-controlled variable photoresistor) for measuring the transmitted light and

vi) an analog or digital meter to display the output from the detector in terms of transmittance or absorbance of sample. In modern colorimeters the filament lamp and filters may be replaced by several light-emitting diodes of different colors.

Q. 5. How many types of colorimeters are there?

Ans.: Generally two types of colorimeters are available: single beam and double beam. Single beam instruments are the simplest and least expensive consisting of a battery-operated tungsten bulb as the light source and a photovoltaic cell as detector. In double beam instruments which are far more expensive, two beams are formed in space by a beam splitter and there are two photodetectors. One beam passes through the reference cell and other through the sample cell simultaneously. The two beams are then analyzed by the respective detectors.

Q. 6. What is the role of filter used in a colorimeter?

Ans.: White light consists of many different wavelengths of light which correspond to different colors. Colored filters help in making the light monochromatic by selecting light of a specific wavelength which is then passed through the test sample. Different chemical substances absorb different frequencies (or wavelengths) of light to different extents. The color of light (or wavelength) which the solute absorbs most is selected in order to maximize the accuracy of the experiment.

Q. 7. What are complimentary colors?

Ans.: Colored solutions absorb light selectively. One or more frequencies are absorbed, the remaining frequencies are transmitted which give solution its color. The absorbed and transmitted colors are said to be complimentary to each other. The selected filter should have the color complimentary to that of the solution. For example if the color of the solution is blue, yellow color is absorbed; if the color of the solution bluish green, red color will be absorbed. So the color of the solution which we see is complimentary to the color absorbed by the sample.

Q. 8. What is the relationship between the solution color, complimentary filter color and the wavelength of transmitted light?

Ans.: Just by looking at the color of the solution we can have an estimate about the complimentary filter color and the wavelength of transmitted light. The following table gives the required relationship.

Table 8. Relationship between solution color, complimentary filter color and filter transmitted wavelength*.

Solution color	Complimentary filter color	Filter transmitted wavelength (nm)
Not visible (colourless)	-	< 400 (Ultraviolet region)
Yellowish green	Violet	400 – 435
Yellow	Blue	435 – 480
Red	Bluish green	490 – 500
Blue	Yellow	580 – 585
Greenish blue	Orange	595 – 610
Bluish green	Red	610 – 750
Not visible	-	> 750 (Infrared region)

* The data in the above Table has been taken from Senior Physical Chemistry by B. D. Khosla, V. C. Garg and Adarsh Khosla. R Chand & Co., New Delhi 16th Edition 2014, p. 112.

Q. 9. What is the wavelength range in which a colorimeter works?

Ans.: The human eye is not capable of seeing radiation with wavelengths outside the visible region of the electromagnetic spectrum. Visible light corresponds to a wavelength range 400-700 nm and a color range from violet to red. This is the wavelength range in which a colorimeter works and therefore, only colored solutions can be studied in colorimeter. If the ultraviolet range is also included, it is generally called a spectrophotometer.

Q. 10. What is the difference between a colorimeter and a spectrophotometer?

Ans.: The major difference between a colorimeter and a spectrophotometer is that a colorimeter uses fixed wavelengths which are in the visible range only but a spectrophotometer can use wavelengths in a wider range which may include ultraviolet (UV) and infrared (IR) also. Colorimeters have comparatively low price, compact size and simple operation whereas spectrophotometers have higher precision, increased versatility but higher cost.

Q. 11. Define absorbance (optical density) and percentage transmittance.

Ans.: Absorbance, $A = \log I_0/I$ is also called the optical density of solution. I_0 and I are the intensity of incident and transmitted light, respectively. Transmittance (T) is defined as the ratio of intensity of transmitted light to intensity of incident light. $T = I/I_0$. This fraction is usually represented as a percentage, called percentage transmittance, $T(\%) = (I/I_0) \times 100$. For example, if all the light passes through a solution without any absorption, then absorbance is zero and percent transmittance is 100. If all the light is absorbed, then percent transmittance is zero and absorption is infinite. If half of the light is absorbed, the intensity of transmitted light is half the intensity of incident light, absorbance, $A = \log 2 = 0.3010$ and percentage transmittance is 50.

Q. 12. How colorimeter is calibrated before use?

Ans.: For accurate results it is always advisable to calibrate the colorimeter before use. Usually a two-step calibration is suggested. We know that a black object absorbs all wavelengths of light and converts them into heat. So a black cuvette is placed in the colorimeter and percentage transmittance is set to zero. Also while measuring the absorbance of a solute in solution, the solvent should not absorb any light so that the light absorbed by the solute only is measured. So the pure solvent is then placed in the cuvette and percentage transmittance is set to 100.

Q. 13. What will be the absorbance and percentage transmittance when 90% of the light is absorbed by the sample?

Ans.: Absorbance, $A = \log I_0/I$. I_0 and I are the intensity of incident and transmitted light, respectively. If the incident light intensity, I_0 is 100, I will be 10 if 90 % of light is absorbed. Under these conditions, absorbance, $A = 1$ and percentage transmittance, $T(\%) = (I/I_0) \times 100 = 10$.

Q. 14. How will you convert percentage transmittance to absorbance and absorbance to percentage transmittance?

Ans.: Absorbance, $A = \log I_0/I$ and percentage transmittance, $T(\%) = (I/I_0) \times 100$. From these definitions of absorbance and percentage transmittance, the following expressions can easily be derived.
Absorbance, $A = 2 - \log T(\%)$ and percentage transmittance, $T(\%) = $ antilog $(2 - $ Absorbance$)$
For example, if percentage transmittance, $T(\%) = 78$, $A = 2 - \log(78) = 0.108$. Similarly if absorbance, $A = 0.349$, $T(\%) = $ antilog $(2 - 0.349) = 44.77$.

Q. 15. State Beer-Lambert law. Define the terms involved.

Ans.: According to Beer-Lambert law, the quantity of light absorbed by a substance dissolved in a fully transmitting solvent is directly proportional to the concentration of the substance and the path length of the light through the solution. Beer-Lambert law is usually stated in mathematical form as follows. Absorbance, $A = \log I_0/I = \varepsilon\, l\, c$, I_0 and I are the intensity of incident light and transmitted light, respectively, ε is called the molar absorptivity or molar extinction coefficient of the sample, l is the path length (the thickness of the colored solution; for example, diameter of a cylindrical cuvette) and c is the concentration of solution in mol/dm^3. If the intensity of transmitted light I, is less than that of the incident light I_0, it is obvious that the sample has absorbed light.

Q. 16. Explain the term molar absorptivity and write its units. How can it be experimentally determined?

Ans.: Molar absorptivity is also known as molar extinction coefficient and is given by $\varepsilon = A/lc$. It is a measure of how well a chemical species absorbs a given wavelength of light. It is characteristic of the nature of solute but does not depend on the concentration of solution. Molar absorptivity does depend on the nature of the solvent, temperature and the wavelength of light used. The SI units of molar absorptivity are m^2/mole but in practice it is usually expressed as dm^3 per mole per centimeter ($dm^3\ mol^{-1}\ cm^{-1}$) or liters per mole per centimeter ($L\ mol^{-1}\ cm^{-1}$).

Since according to Beer-Lambert law, $A = \varepsilon lc$, A versus c plot should be linear passing through origin since the path length l is a constant quantity. From the slope of the plot molar absorptivity, ε can be determined.

Q. 17. What are the conditions for validity of Beer-Lambert law.

Ans.: According to Beer-Lambert law, absorbance, $A = \log I_0/I = \varepsilon lc$. The following are the necessary conditions for validity of the law. i) The light passed should be monochromatic. This is because the simple detector (photoresistor) used in colorimeter is designed only for monochromatic light. ii) The law is valid only for dilute solutions. There can be deviations in absorptivity coefficients at high concentrations (> 0.01M) due to change in the charge distribution of molecules in close proximity. At higher concentrations the refractive index of solution also varies considerably. iii) The colored medium must not scatter the radiation and therefore there should be no turbidity in the solution. iv) Deviations are also observed if the dissolved entity dissociates, associates or reacts with the solvent to form a product having different absorption.

Q. 18. How Beer-Lambert law is verified?

Ans.: According to Beer-Lambert law, absorbance varies linearly with the concentration of solution. Measured absorbance of colored solutions of different concentrations at a fixed wavelength is plotted against the concentration of solution. The law is verified if a linear plot passing through origin is obtained.

Q. 19. Why aqueous solutions of potassium dichromate do not obey Beer-Lambert law?

Ans.: Aqueous solutions of potassium dichromate do not obey Beer-Lambert law due to hydrolysis of dichromate ions. In aqueous solutions dichromate ions react with water to form chromate ions according to the reaction

$$Cr_2O_7^{2-} + H_2O \rightleftharpoons 2\ CrO_4^{2-} + 2\ H^+$$

However, if the solution is prepared in acidic medium, the hydrogen ions in acid shift the equilibrium to the left and the salt remains practically unhydrolyzed. Therefore, potassium dichromate solutions obey Beer-Lambert law in acidic medium.

Q. 20. Which parameter determines the intensity of color?

Ans.: According to Beer-Lambert law, absorbance is directly proportional to the concentration of colored solution. Thus the parameter which determines the intensity of color is the absorbance of solution. More intense the color more is the absorbance at a given wavelength.

Q. 21. Which parameter can be used to identify the compound?

Ans.: Different chemical substances absorb different frequencies of light. Molar absorptivity (ε) is different for different substances and this is the parameter which is used to identify compound. Molar absorptivity is a function of the nature of solute; it does not depend on the concentration of solution. The specific values for different chemical species at specified wavelengths of light can be found in chemical reference manuals.

Q. 22. How the experimentally determined absorbance data can be used to determine the concentration of colored solution?

Ans.: The experimentally determined absorbance data can be used to determine the concentration of colored solution (test solution) by the application of Beer-Lambert law. Absorbance of a set of solutions of known concentration is determined and plotted against the respective concentrations to obtain a linear calibration plot passing through origin. The calibration plot is then used to determine the unknown concentration from the experimentally determined absorbance of the test solution. The unknown concentration can also be determined from Beer-Lambert law without a calibration plot if for a given sample the molar absorptivity and path length are known.

Q. 23. Molar absorptivity of a given sample is 1.5 L mol^{-1} cm^{-1}. If measured absorbance using a cuvette of path length 1 cm is 0.450, what is the concentration of solution?

Ans.: According to Beer-Lambert law, absorbance, $A = \log I_0/I = \varepsilon l c$. $0.450 = 1.5$ L mol^{-1} cm^{-1} × 1 cm × c, Concentration of solution, c = 0.3 mol L^{-1}.

Q. 24. A solution of thickness 2 cm transmits 60% of incident light. What is the concentration of solution if molar absorptivity of substance is 500 dm^3 mol^{-1} cm^{-1}.

Ans.: Absorbance, $A = 2 - \log (T (\%)) = 2 - \log 60 = 0.222$. According to Beer-Lambert law, absorbance, $A = \varepsilon l c$. Concentration, c = 0.222/(500 × 2) = 2.22 × 10^{-4} mol dm^{-3}.

Q. 25. The measured absorbance of a solution A with concentration of 0.20 M is 0.582. Absorbance of another solution B of the same substance measured under same conditions is 0.795. What is the concentration of solution B?

Ans.: According to Beer-Lambert law, absorbance, $A = \varepsilon l c$. For solution of same substance under same conditions, ratio of absorbance is equal to the ratio of concentrations. If subscript 1 refers to the solution of unknown concentration and subscript 2 to the solution of known concentration, $c_1/c_2 = A_1/A_2$ and unknown concentration $c_1 = (0.795/0.582) \times 0.20 = 0.27$ M.

Q. 26. Can colorimetric method be used for a system which does not obey Beer-Lambert law?

Ans.: The use of a photoelectric colorimeter is not limited to the condition of Beer-Lambert law being obeyed. A calibration curve even when non-linear can be used for estimating concentrations by interpolation as in other physical methods of analysis.

Q. 27. What types of system are suitable for colorimetric method of analysis?

Ans.: Colorimetry is one of the most useful and widely used tools available for quantitative analysis due to relatively inexpensive instrumentation required and ease of data analysis. For example i) any organic compound containing one or more chromophores can easily be studied, ii) a number of inorganic species also absorb light and can be directly determined, iii) non-absorbing species can also be studied by selective reaction with certain reagents to form colored complexes. Moreover, colorimetry can also be used for the determination of dissociation constant of indicators as well as finding the composition of a complex.

Q. 28. What is the principle of Job's method for determination of composition of a complex?

Ans.: Job's method, also called the method of continuous variations, is used in analytical chemistry to find the stoichiometry of a complex. In this method, the total number of moles of reactants is kept constant but the mole ratio or mole fraction of reactants is varied for a series of measurements. An extensive property which depends on the amount of reactant consumed or product formed in the reaction is measured. This property may be the color/color intensity of a reactant or product, the mass of a precipitate formed, the amount of heat produced etc. The maximum change in property will occur when the mole ratio of the reactants is closest to the actual stoichiometric mole ratio in the chemical equation.

When one of the reactants or products is colored, a colorimeter can be used to study the reaction. From the solutions prepared with different mole ratio of reactants, the one with most intense color is used for selecting the wavelength for absorbance measurements. The absorbance of this solution is determined at different wavelengths and the wavelength where maximum absorbance is obtained is selected. Absorbance measurements at the selected wavelength are used to monitor the color intensity of the series of reactant mixtures prepared at different mole ratio of reactants. If for example, the colour intensity of products is more than that of the reactants, the maximum in the absorbance versus mole ratio plot corresponds to the composition of the complex. If on the other hand, the colour intensity of reactants is more than that of the products, the plot will show a minimum corresponding to the stoichiometry of the reaction.

Q. 29. What is the necessary condition for colorimetric method to be used for determination of composition of a complex using Job's method?

Ans.: Colorimetric method can be used for the determination of composition a complex using Job's method only when at least one of the reactants or products is colored.

Q. 30. Give an example of a reaction where composition of a complex can be determined colorimetrically.

Ans.: A common example is ferric salicylate complex formed by reaction between ferric ions and salicylic acid. The ferric ions for this reaction are provided by ferric ammonium sulphate ($NH_4Fe(SO_4)_2.12\ H_2O$), also called ferric alum. The composition of this complex can be determined colorimetrically since the reactants are colorless but the complex is deep violet in color. The wavelength at which the absorbance of the complex is maximum can be selected by choosing an appropriate filter. The absorbance at the selected wavelength will be proportional to the concentration of complex according to Beer-Lambert law. As discussed in the previous question, the concentration of the complex will be maximum when reactants are present in stoichiometric amounts; that is in the same mole ratio as required for the complex formation. A set of solutions is prepared by mixing equimolar solutions of reactants (0.001M each) in varying proportions. The absorbance values, determined at the selected wavelength are plotted against composition of the reaction mixture. The maximum in the absorbance versus composition plot corresponds to the composition of the complex.

Q. 31. Under what conditions ferric salicylate complex is stable?

Ans.: Ferric salicylate complex is stable in acidic medium (pH 2.6 – 2.8). Due to this reason ferric alum and salicylic acid solutions are always prepared in dilute hydrochloric acid solution (0.002 M).

Q. 32. Give some other examples of colored complexes which can be studied colorimetrically.

Ans.: Examples of other colored complexes include the following. i) Ferric thiocyanate complex formed by reaction between ferric nitrate and potassium thiocyanate is blood red in color, ii) Nickel-phenanthroline complex formed by reaction between nickel nitrate and 1,10 phenanthroline is pink in color and iii) Ferroin which is a complex of ferric ions and 1,10-phenanthroline is also red in color.

Q. 33. Explain how dissociation constant of phenolphthalein can be determined colorimetrically?

Ans.: Phenolphthalein is a weak acid, the ionized form of which is colored pink and the unionized form is colorless. In basic medium, the neutralization of hydrogen ions produced by the ionization of indicator pushes the equilibrium to the right and the indicator exists in predominantly ionized form. The following equation can easily be derived from the ionization equilibrium of the indicator (Acid-base Indicators Q. 11).

$$pH = pK_{In} + \log \frac{[In^-]}{[HIn]} = pK_{In} + \log \frac{[Ionized]}{[Unionized]}$$

It is apparent from this equation that with increase in pH of solution, the degree of ionization of indicator also increases since pK_{In} is constant. The color intensity and therefore, the absorbance of the solution also increases with increase in pH because the ionized form is colored. According to Beer-Lambert law, absorbance $A = \varepsilon cl = \varepsilon [In^-] l$, since only the ionized form is colored. In alkaline medium at pH > 10, it is safe to assume that the whole of indicator exists in ionized form and the absorbance becomes maximum, A_{max}. At this stage $[In^-]$ = the total indicator concentration, [HIn] and according to Beer-Lambert law, absorbance $A = A_{max} = \varepsilon [HIn] l$. Therefore, in the above equation,

$$\log \frac{[In^-]}{[HIn]} = \log \left(\frac{A}{A_{max}}\right) \text{ and } pH = pK_{In} + \log \left(\frac{A}{A_{max}}\right) \text{ or } \log \left(\frac{A}{A_{max}}\right) = pH - pK_{In}$$

Absorbance, A is determined at different pH values and log A/A_{max} is plotted against pH. The slope of the straight line plot should be unity and intercept is equal to $-pK_{In}$. The dissociation constant of phenolphthalein (K_{In}) can be determined from its pK_{In} value.

Q. 34. What are photometric titrations?

Ans.: A photometric titration is a titration which is accompanied by change in the color/color intensity of the reaction mixture. The change in color/color intensity may either be due to consumption of colored reactant or formation of colored product. The reaction is followed by measuring the color intensity of the reaction mixture. The method involves plotting measured absorbance of reaction mixture against volume of the titrant added and analyzing data.

Q. 35. Describe briefly various types of photometric titrations.

Ans.: Photometric titrations can be broadly classified into four types.
    a) Titrant alone absorbs: Initially the absorbance of the reaction mixture remains unchanged since the added titrant is being consumed but there is a sudden rise in absorbance after the equivalence point when titrant is present in excess. When the measured absorbance of reaction mixture is plotted against the volume of the titrant added, the point of intersection of the two lines drawn through the data corresponds to the equivalence point.

For example, bromine formed by the reduction of bromate ions in the presence of alkali metal bromide in acidic medium reacts with arsenic (III) to form arsenic tribromide. When arsenic is titrated with bromate-bromide in acidic medium, the absorbance is measured in the visible range at wavelength where the titrant bromine absorbs.

b) The product of the reaction absorbs: Absorbance increases as more and more of the absorbing product is formed but remains constant after the equivalence point since the product concentration does not increase thereafter. A common example is the titration of copper (II) with EDTA. Both copper (II) and copper (II)-EDTA complex absorb at 625 nm but the molar absorptivity of the copper (II)-EDTA complex is much higher. So as more and more of copper-EDTA complex is formed, absorbance increases and becomes constant after the equivalence point since the titrant EDTA does not absorb at this wavelength.

c) Absorbing reactant is converted into non-absorbing product: In this case when an absorbing reactant is titrated with another non-absorbing reactant, the concentration of absorbing reactant decreases as the reaction progresses while that of the non-absorbing product increases and therefore, absorbance shows a decrease till the equivalence point is reached. Thereafter the absorbance remains constant since the other reactant which is used as titrant is non-absorbing. For example, titration of thiosulphate ion with triiodide ion involves reduction of colored triiodide ion to colorless iodide ion while thiosulphate ion is converted to tetrathionate ion. Both thiosulphate and tetrathionate ions are non-absorbing.

d) A colored species is converted to a colorless product by a colored titrant: In this case initially absorbance decreases due to consumption of colored reactants and formation of colorless product but increases after the equivalence point due to the presence of excess of colored titrant. For example in the bromination of a red dyestuff, bromine in solution is reddish-brown. When free bromine reacts with the dye, the solution becomes colorless. The reaction can be monitored by measuring the absorbance of the reaction mixture.

Q. 36. How experimental data in photometric titrations is corrected for dilution of reaction mixture by added titrant.

Ans.: In any colorimetric titration, the added titrant dilutes the reaction mixture and this dilution effect causes the straight line segments to curve. The effect of dilution is rendered negligible either by using a concentrated solution of titrant so that lesser volume is used or by applying a correction to the absorbance value. To apply the correction factor, absorbance is multiplied by $[(V + v)/V]$ where V is the volume of the solution being titrated and v that of the titrant added. Good linear plots can be obtained by applying this correction to the absorbance data.

Q. 37. What is an absorption spectrum?

Ans.: When electromagnetic radiation is passed through a sample, some of its energy may be absorbed by the sample. Absorption spectrum shows how the intensity of absorption varies with the frequency of radiation. Usually it is a plot of the absorbance or transmittance against the frequency or wavelength of incident light.

Q. 38. What are the applications of colorimetry in analytical chemistry?

Ans.: Colorimeric methods are widely used in analytical chemistry in the estimation of metal ions in solution. Most colored metal ions can easily be analyzed by this method. For example, the characteristic color of the permanganate ion can be used in the colorimetric analysis for Mn. However, many metal ions do not show strong colors, particularly at low concentrations. In such cases, highly colored complexes can be formed between metal ions and organic or inorganic complexing agents. The necessary condition is that the colored complex formed should have high absorptivity and minor changes in pH or temperature should not affect colour. For example iron (II) forms intense red color complex with 2, 2´-bipyridine which may be exploited to determine iron concentrations in the ppm range.

Q. 39. What are the applications of colorimetry in medicine?

Ans.: Colorimetric assays have a wide spectrum of applications in drug analysis and biomedical analytical chemistry. For example, the colorimetric assay of sulphadiazine is based on the acid-catalyzed equilibrium reaction that occurs between vanillin (an aldehyde) and sulphadiazine (an arylamine). The product is a schiff's base and is yellow in color. As an example of application in biomedical analytical chemistry, cholesterol interacts with glacial acetic acid and acetic anhydride to form a colored product whose absorption at 630 nm is directly proportional to the level of cholesterol present in serum.

Q. 40. How do you compare colorimetric method with the titration method for estimating the concentration of substance?

Ans.: Both titration method and colorimetric method used to determine the unknown quantity of a substance, are based on the observation of colour changes in the reaction mixture. However, the mechanism responsible for color change is different in each case. Colorimetry, a more direct technique for determining concentration, measures how much light is actually absorbed by the coloured sample. The measured absorbance is directly proportional to the concentration of analyte. Titration by contrast is an indirect method which relates the volume of reagent added to concentration. Colorimetry requires the use of an instrument called a colorimeter while titration requires only reagents and glassware. Finally, titration requires a reagent that can react with the substance to be analyzed whereas colorimetry requires that the reactant or the product should be able to absorb light in the visible region of the spectrum. Each technique has its own limitations. For titration method to work a suitable indicator should also be available and the substance to be analyzed must react quickly and completely with the indicator. Titration may also prove problematic if multiple chemicals in the sample can react with the added titrant. For example adding a base to a sample that contains multiple acids would not tell how much of each acid is present. Colorimetry by contrast, can also be used if the sample contains more than one compound provided they absorb at different wavelengths in the visible region of the electromagnetic spectrum. Colorimetric method however, requires that the concentration of the absorbing substance must be within a given range. If the concentration is too small, the amount of light absorbed will be so small it will be undetectable. If the concentration is too high, the absorbance will cease to increase with increasing concentration.

11. Liquid state – Density

Q. 1. What is density?

Ans.: Density is a physical property defined as mass per unit volume. Regardless of the sample size density is always constant. Density is an intensive property which tells us how heavy or light a particular object is.

Q. 2. Why different substances have different densities?

Ans.: Density varies from substance to substance due to differences in the relation of mass and volume. Same mass of different substances can occupy widely different volumes. For example one kilogram of mercury occupies much less volume as compared to one kilogram of water. Density of water is approximately 1 g/mL whereas that of mercury is 13.6 g/mL. Mercury is 13.6 times denser than water.

Q. 3. What are SI units of density? What is the density of water in SI units?

Ans.: The SI units for density are kilogram per cubic meter (kg/m^3). The density of water, commonly expressed as $1 g/cm^3$, is equal to $1000 kg/m^3$ in SI units. However, density is temperature-dependent and the value is 1.00 g/cm^3 at 4.0°C. It decreases slightly as the temperature is increased.

Q. 4. The density of a substance is 1.25 g/mL. What is the mass of 0.50 litres of substance in grams?

Ans.: Density of substance = 1.25 g/mL. 1 mL of substance has mass = 1.25 g.
1 L of substance has mass = 1.25×10^3 g. 0.50 L of substance has mass = $1.25 \times 10^3 \times 0.50$ g = 625 g.

Q. 5. The density of a substance is 2.80 g/mL. What volume does 2 kg of the substance occupy?

Ans.: Density of substance = 2.80 g/mL. 2.80 g of substance occupy volume = 1 mL.
2 kg of substance occupy volume = $(1/2.80) \times (2 \times 10^3)$ mL = 714 mL.

Q. 6. Two liquids A and B have densities 1.50 and 3.75 g/mL, respectively. When both the liquids are poured in a container which liquid will float on the top of the other?

Ans.: The lighter liquid will float on the top of the denser liquid. Liquid A with lower density is lighter than liquid B which has higher density. Therefore, liquid A will float on the top of the liquid B.

Q. 7. Which are the most dense and least dense elements in periodic table?

Ans.: In the periodic table, hydrogen is the least dense element while osmium is the most dense element.

Q. 8. What is the effect of temperature on density of a substance?

Ans.: Density generally decreases with increase in temperature since volume increases with increase in temperature.

Q. 9. Why ice is less dense than water?

Ans.: Density of water also increases with decrease in temperature but ice is less dense than water. This is unusual as solids are generally denser than their liquid counterparts. Ice is less dense than water due to hydrogen bonding. In the water molecule, the hydrogen bonds are strong and compact. As the water freezes into the hexagonal crystals of ice, these hydrogen bonds are forced farther apart and the resulting increase in volume is responsible for the decrease in density of ice. This explains why ice is less dense and therefore, floats on water.

Q. 10. Why hot air balloons fly high?

Ans.: The hot air inside the balloon is less dense than the cold air outside. The hot air balloon is therefore lighter than the surrounding air in the atmosphere and so it rises.

Q. 11. What is the effect of pressure on density?

Ans.: Density of a gas increases with increase in pressure because density is defined as mass per unit volume and volume decreases as pressure increases. The decrease in volume with increase in pressure is explained in Boyle's Law according to which at constant temperature, $P_1V_1=P_2V_2$, where P = pressure and V = volume. Since liquids and solids are generally considered as incompressible their density is not much affected by increase of pressure.

Q. 12. What is the relation of Archimedes principle to density?

Ans.: When an object is immersed in water or another liquid, it is subjected to an upward buoyant force from the surrounding fluid due to greater fluid pressure below the body than above it. Due to this upward buoyant force the apparent weight of an object immersed in a liquid decreases by an amount equal to the weight of the volume of the liquid that it displaces. This is Archimedes' Principle. Since 1 mL of water has a mass almost equal to 1g, if the object is immersed in water, the decrease in the weight of the object (in grams) will equal (almost exactly) the volume (in mL) of the object weighed. Knowing the mass and the volume of an object allows us to calculate the density.

Q. 13. An object weighs 36 g in air and has a volume of 8.0 cm³. What will be its apparent weight when immersed in water?

Ans.: According to the Archimedes principle the upward buoyant force acting on the body is equal to the weight of the fluid it displaces. The object displaces fluid equal to its own volume. Therefore, the volume of

water displaced = 8 cm³. Taking the density of water as unity, the upward (buoyancy) force is 8 g.
The apparent weight will be (36 g) – (8 g) = 28 g.

Q. 14. A piece of metal weighs 10.50 g in air and 9.20 g in water. What is the density of the metal?

Ans.: When immersed in water, the metal object displaces (10.50 – 9.20) g = 1.30 g of water whose volume is (1.30 g)/(1.00 g cm^{-3}) = 1.30 cm³. Since the object displaces fluid equal to its own volume, the volume of the metal is also 1.30 cm³. The density of the metal is thus (10.50 g)/(1.30 cm³) = 8.08 g cm^{-3}.

Q. 15. What is the density of air?

Ans.: According to International Standard Atmosphere (ITA), the density of air is 1.225 kg/m³ at sea level at 15°C. IUPAC value for density of dry air at standard temperature and pressure (0°C and 100 kPa) is 1.2754 kg/m³. At 20°C and 101.325 kPa (1 atm), the density of dry air is 1.2041 kg/m³. However, these standard values are only an approximation since atmospheric air always contains some moisture and density is also affected by the amount of water vapour in air.

12. Liquid state - Surface tension

Q. 1. Why a liquid drop is spherical?

Ans.: A liquid drop is spherical in shape due to surface tension forces. The cohesive forces among liquid molecules are responsible for the phenomenon of surface tension. The inward forces on surface molecules tend to contract the surface and spherical shape has minimum surface to volume ratio.

Q. 2. Why surface molecules have high energy?

Ans.: We know that attractive interactions lead to lowering of energy. In the bulk of the liquid, each molecule is pulled equally in every direction by neighboring liquid molecules, resulting in a net force of zero whereas surface molecules are attracted only by molecules below in the bulk of liquid and therefore they have higher energy as compared to the molecules in the bulk.

Q. 3. Define surface tension and surface energy. What are their units?

Ans.: The phenomenon of surface tension is due to the cohesive forces in liquids which pull the surface molecules into the bulk thereby minimizing surface area. It is usually defined as the force acting perpendicular to a line of unit length on the surface. Surface energy is defined as the work that is required to increase the area of the surface of a liquid by one centimeter square at constant temperature, pressure and composition. The terms surface tension and surface energy convey the same meaning. Surface tension is defined as force per unit length (N/m) while surface energy is defined as energy per unit area of the surface (J/m^2). The two units are equivalent since J = Nm. Due to the high energy hydrogen bonding interactions between water molecules, surface tension of water is higher than most other liquids. Water has surface tension 0.072 N/m and surface energy 0.072 J/m^2 at room temperature.

Q. 4. What are various methods used for the experimental determination of surface tension?

Ans.: There are several methods of measuring surface tension. These include capillary rise method, stalagmometer method, Wilhelmy plate or duNouy ring method, maximum bubble pressure method and methods analyzing shape/size of the hanging liquid drop. The choice of the method usually depends on whether the primary consideration is extreme absolute accuracy, practical convenience, speed of working, sample size etc. The two most common and inexpensive methods used in the undergraduate chemistry laboratory are capillary rise method and stalagmometer method.

Q. 5. What is the principle of capillary rise method for determination of surface tension?

Ans.: In the capillary rise method, a capillary tube is dipped in the liquid. If the adhesion forces that is, the interaction forces between the liquid molecules and the capillary walls are stronger than the forces of cohesion between the liquid molecules, the liquid wets the walls and rises in the capillary tube. On the other hand for a non-wetting liquid, stronger cohesion forces between liquid molecules as compared to the adhesion forces between the liquid molecules and the capillary walls cause decrease of the liquid level in the capillary below that in the chamber. If the capillary is circular in cross-section and its radius is sufficiently small, then the meniscus will be approximately semispherical with radius equal to the radius of the capillary. The force due to the surface tension acting along the perimeter of the meniscus is given by $f_1 = 2\pi r \gamma \cos\theta$. In the expression for f_1, r is the capillary radius, γ the surface tension of the liquid and θ is the wetting contact angle. Contact angle, θ is the angle which the tangent to the liquid surface makes with wall of the capillary tube. The force f_1 is balanced by the force of gravity, f_2 due to the mass of the liquid raised in the capillary to the height h. In the case of non-wetting liquid, it is lowered to a distance minus h. Force $f_2 = \pi r^2 h d g$, where d is the liquid density and g the acceleration due to gravity. At equilibrium, the liquid does not move in the capillary, $f_1 = f_2$, and hence $2\pi r \gamma \cos\theta = \pi r^2 h d g$ and $\gamma = rhdg/2\cos\theta$. This general expression can be used to determine surface tension of a liquid from the experimentally measured value of height h to which the liquid rises in the capillary tube and known values of radius of capillary r and contact angle θ. The expression simplifies to $\gamma = rhdg/2$ for a liquid which wets the capillary walls completely; the contact angle θ is assumed to be 0° and $\cos\theta = 1$. For a liquid such as mercury which does not wet the walls of glass capillary, it is assumed that contact angle, $\theta = 180°$ and $\cos\theta = -1$. As the meniscus is lowered by the height, - h, in this case also the calculated surface tension will be positive.

Q. 6. A liquid of surface tension 9 x 10⁻² Nm⁻¹ and density 0.90 g/mL rises to a height of 1 cm in a capillary tube of radius 2 mm. Calculate contact angle, θ.

Ans.:
$$\gamma = \frac{rhdg}{2\cos\theta}, \quad \cos\theta = \frac{rhdg}{2\gamma} = \frac{2 \times 10^{-3} \times 1 \times 10^{-2} \times 900 \times 9.81}{2 \times 9 \times 10^{-2}} = 0.981, \theta = 11.19°$$

All data has been converted to SI units. Surface Tension, $\gamma = 9 \times 10^{-2}$ Nm⁻¹, Density, $d = 0.90 \times 10^3$ g/mL = 900 kg/m³, Height, $h = 1 \times 10^{-2}$ m, Radius of capillary, $r = 2 \times 10^{-3}$ m.

Q. 7. What is the principle of stalagmometer method for determination of surface tension?

Ans.: Stalagmometer method is the simplest and most commonly used method for routine measurement of surface tension in undergraduate physical chemistry labs. Stalagmometer consists of a glass tube with a bulb in the middle section and a capillary tube in the lower end. The narrow capillary tube in the lower end enables the liquid to fall out of the tube as drops. A rubber tube with a pinch cock is attached to the upper end of the tube. Two reference marks, one above the bulb and another below the bulb are etched on the **stalagmometer** and it is clamped in vertical position on a stand. The rate of flow of drops can be adjusted with the help of the pinch cock. A drop of liquid, formed at the tip of the capillary tube, slowly increases in size and falls off when the weight of the hanging drop reaches a maximum value which is dependent on the characteristics of the liquid and the dimensions of the capillary tube. The drop falls when the weight (mg) is equal to the circumference ($2\pi r$) multiplied by the surface tension (γ). Based on Tate's law, $mg = 2\pi r\gamma$. The surface tension can be calculated provided the radius of the tube (r) and mass of the fluid drop (m) are known. For a liquid which wets the capillary tip, r is taken as outer radius of capillary and for liquids which do not wet the capillary tip, r is taken as the inner radius of the capillary. Alternatively, since the surface tension is proportional to the mass of the falling drop, the fluid of interest may be compared to a reference fluid of known surface tension (typically water). Since determination of the radius of the capillary is cumbersome, it is usually used as a relative method. For a given capillary, all other terms cancel out and the ratio of surface tensions of the test liquid and the reference liquid is equal to the corresponding ratio of mass of drops, $\gamma_1/\gamma_2 = m_1/m_2$, the subscripts 1 and 2 refer to the test liquid and the reference liquid, respectively. A fixed number of drops (say 10 or 20) of the test liquid and the reference liquid are collected in a pre-weighed container and the weight per drop is determined. This is also called drop weight method.

The precision of the method can be increased by applying a correction factor, f. In actual practice when a drop falls, some liquid remains attached to the tip and the correction factor, f takes care of that. Tate's law expression can then be written as $mg = 2\pi r \gamma f$. The correction factor, f is a function of the dimensionless ratio, $r/v^{1/3}$, where v is the drop volume. The volume of drop v can be determined from the mass of the drop and the density of the liquid. Correction factor, f for a range of $r/v^{1/3}$ values is also available in literature.

Q. 8. What precautions are required in the determination of surface tension by stalagmometer method.

Ans.: In order to get accurate data it is necessary that the stalagmometer should be clamped vertical. The rate of fall of liquid drops should be adjusted to about 10 – 15 drops per minute. If the rate of fall of drops is higher, the drops will not be properly formed and if it is lower than this then the time taken to complete the experiment will increase considerably. Since surface tension is temperature-dependent, the stalagmometer should preferably be placed in a constant temperature bath.

Q. 9. How drop number method can be used to determine the surface tension of a liquid?

Ans.: Drop weight method for the determination of surface tension using a stalagmometer is described in detail in Q. 7. Drop number method can simply be derived from drop weight method. Since mass (m) = volume (v) x density (d), the ratio of mass of test liquid and reference liquid can be written as $m_1/m_2 = v_1 d_1/v_2 d_2 = \gamma_1/\gamma_2$ where v is the volume of each drop and d is the density of liquid. The subscripts 1 and 2 refer to the test liquid and reference liquid, respectively. Now volume of one drop is equal to the total volume divided by the number of drops. If V_1 and V_2 are the total volumes and n_1 and n_2 are the number of drops of test liquid and reference liquid, respectively then the volume per drop, $v_1 = V_1/n_1$ and $v_2 = V_2/n_2$. If the drops are counted between two fixed marks in the stalagmometer, $V_1 = V_2 = V$, the volume of liquid between the two marks and $v_1/v_2 = n_2/n_1$. Therefore, the ratio of surface tensions, $\gamma_1/\gamma_2 = n_2 d_1/n_1 d_2$. The densities of the test and reference solutions can be experimentally determined or obtained from the literature. Thus just by counting the number of drops formed from a fixed volume of the test liquid and reference liquid, surface tension of the test solution can be determined.

Q. 10. Calculate the radius of a water drop falling from a capillary tube 1.00 mm in diameter. Surface tension of water = 7.00 x 10⁻² Jm⁻².

Ans.: Just before the drop falls, the surface tension force (= $2\pi r \gamma$) holding it up equals the force of gravity (= mg) pulling it down. Assuming the drop to be spherical

$2\pi r \gamma = mg = 4/3 \pi R^3 dg$ where r is the radius of capillary, R is the radius of liquid drop and d is the density of water.

Radius of the drop,

$$R = \left(\frac{3r\gamma}{2dg}\right)^{1/3} = \left(\frac{3 \times 0.50 \times 10^{-3} \times 7 \times 10^{-2}}{2 \times 1000 \times 9.81}\right)^{1/3} = 1.75 \times 10^{-3} \, mm.$$

Radius of capillary, R = ½ diameter = 0.50 mm = 0.50 x 10^{-3} m, Surface tension, γ = 7.00 x 10^{-2} Jm^{-2}, Density of water, d = 1 g cm^{-3} = 1000 kg m^{-3}, Acceleration due to gravity, g = 9.81 ms^{-2}.

Q. 11. State and explain Young and Laplace equation in surface chemistry.

Ans.: There exists a pressure difference across a curved liquid-air interface or we can say that the interface is curved due to the pressure difference on the two sides of interface. Pressure is always more on the concave side of the meniscus. In the case of a wetting liquid with concave meniscus for example, the liquid rises in the capillary tube till the mechanical equilibrium is attained when the pressure difference, ΔP is balanced by the hydrostatic pressure caused by the raised liquid in the capillary. According to Young and Laplace equation the pressure difference, ΔP is related to the surface tension of liquid by the equation, ΔP = 2γ/r. For a narrow capillary of small radius, the interface is approximately semispherical and r the radius of curved meniscus is approximately equal to the radius of the capillary tube. Similar considerations can be made for a convex meniscus. Young and Laplace equation can also be used to derive the expression for surface tension determination.

Q. 12. What is the difference between surface tension and interfacial tension?

Ans.: Surface tension and interfacial tension are essentially the same. Both refer to the interface between two phases. Generally when one of the phases is air it is called surface tension. Surface tension defined for a single liquid always refers to the liquid-air interface whereas interfacial tension is defined for the interface between two different phases such as two immiscible liquids.

Q. 13. What is the effect of temperature on surface tension?

Ans.: Due to increase in the thermal motion of molecules at higher temperatures, intermolecular interactions, i.e. the cohesive forces between molecules decrease with increase in temperature and therefore, in general surface tension decreases with increase in temperature. Thus surface tension measurements should be performed at constant temperature and temperature must be stated while reporting surface tension data. If due to non-availability of constant temperature bath in undergraduate laboratory, measurements are done at room temperature, room temperature should be noted at the start and end of the experiment and the average of these two temperatures should be stated while reporting the result.

Q. 14. What is an amphiphilic substance?

Ans.: A typical amphiphilic molecule consists of two parts, a polar head group and an essentially non-polar or hydrophobic group which is in general a single or branched chain hydrocarbon. Because of its dual affinity, some amphiphilic molecules exhibit surface activity that is they exhibit a very strong tendency to migrate to the surface and to orientate so that the polar groups lie in water or other polar solvent and the non-polar or hydrophobic groups are placed out of it.

Q. 15. What are surface active substances? Are all amphiphilic substances surface active?

Ans.: Surface active substances are also called surfactants. Surfactants are amphiphilic molecules which have a tendency to migrate to the liquid-air interface thereby lowering the surface tension of the liquid in which they are dissolved. However, all amphiphilic substances are not surface active. Only those amphiphiles are surface active in which the hydrophilic and hydrophobic tendencies are more or less balanced. If the amphiphilic molecule is too hydrophilic or too hydrophobic, it is likely to stay in one of the phases and does not exhibit surface activity. Thus all surfactants are amphiphilic substances but all amphiphilic substances are not surfactants. Common examples of surfactants include salts of fatty acids (soaps), alkylbenzene sulphonates (detergents), sodium lauryl sulphate (foaming agent).

Q. 16. What is hydrophile-lipophile balance (HLB)? What is the significance of HLB value of a surfactant?

Ans.: Hydrophile-lipophile balance, also called HLB value, is a measure of the degree to which a surfactant molecule is hydrophilic or lipophilic (hydrophobic). Each surfactant is assigned a numerical value based on the size and strength of hydrophilic and lipophilic moieties present. HLB scale ranges from 0 to 20; 0 corresponds to completely lipophilic/hydrophobic molecule and 20 corresponds to a completely hydrophilic/lipophobic molecule. In other words, low HLB surfactants are more lipophilic and exhibit more oil solubility while high HLB surfactants are more hydrophilic and exhibit more water solubility. It is obvious that amphiphiles with very low and very high HLB value will not be surface active.

Q. 17. Why n-amyl alcohol is surface active while methyl alcohol is not?

Ans.: As already explained in previous questions, for a substance to be surface active there must be a balance between the hydrophilic and hydrophobic parts of the molecule. Although both amyl alcohol and methyl alcohol contain hydrophilic hydroxyl group, methyl alcohol is not surface active due to the very weak hydrophobicity of methyl group. Methyl alcohol is predominantly hydrophilic in nature. On the other hand, the hydrophobicity of hydrocarbon chain in n-amyl alcohol ($CH_3(CH_2)_4 OH$, pentane-1-ol) with five carbon atoms is sufficient to balance the hydrophilic hydroxyl group and therefore, amyl alcohol is surface active.

Q. 18. What will be the orientation of amyl alcohol molecules on the surface of water?

Ans.: When a surface active substance is dissolved in water the water-insoluble hydrophobic group extends out of the bulk water phase into the air or into the oil phase while the water-soluble head group remains in the water phase. Therefore, amyl alcohol molecules will be oriented vertically in such a way that the hydroxyl groups are in water phase while the hydrocarbon chains extend out of water into air.

Q. 19. What is micellization? Define critical micellar concentration?

Ans.: Micellization is the process of aggregation of surfactant molecules in solution to form micelles. In a polar solvent such as water, the micellar structures are generally spherical in shape and are formed in such a way that the hydrophilic head groups remain outside while hydrophobic tails coil inside the micelle. Due to this aggregation behavior of surfactants, they are also termed as association colloids, colloidal surfactants or micellar systems. Micelles form only above a certain minimum concentration of surfactant. The surfactant concentration above which micellization starts is known as the critical micellar concentration (cmc) of surfactant.

Q. 20. What are reverse micelles?

Ans.: In non-polar medium the micellar structures are reversed. The hydrophobic tails remain outside in non-polar solvent while hydrophilic head groups coil inside the micelle. Such systems are called reverse micelles and have the capability of solubilizing polar solutes in non-polar medium.

Q. 21. What is the difference between a soap and detergent?

Ans.: Soaps are usually metal salts of long chain fatty acids and are prepared from vegetable oils and animal fats. They cannot be used effectively in hard water due to the formation of insoluble precipitates. Detergents are sodium salts of long chain hydrocarbons like alkyl sulphates and alkylbenzene sulphonates and are prepared from hydrocarbons. Detergents are effective in soft, hard and salt water. Detergents are generally prepared synthetically and are non-biodegradable while soaps are prepared from natural resources and are biodegradable.

Q. 22. How surfactants are classified?

Ans.: Most surfactant tails are very similar consisting of a hydrocarbon chain which can be linear, branched or aromatic. Surfactants are usually classified on the basis of head groups as anionic, cationic and non-ionic surfactants. A surfactant is anionic if the head group carries a negative charge (e.g., carboxylates, sulphonates, alkylbenzene sulphonates), cationic if the head group carries a positive charge (e.g., quaternary ammonium salts) and zwitterionic or amphoteric if the head group carries both positively and negatively charged groups (e.g., betaines such as cocamidopropyl betaine, lauryl betaine, coco betaine). The head group of a non-ionic surfactant is also hydrophilic but it does not does not have any charged groups (e.g., ethoxylated aliphatic alcohols, polyoxyethylene derivatives, polyoxyethylene glycol esters).

Q. 23. What are the major uses of different types of surfactants?

Ans.: Surfactants are used as emulsifiers, detergents, dispersing agents, foaming agents and wetting agents due to their ability to lower surface tension of the liquid in which they are dissolved. Moreover, since surfactants aggregate in solution to form micelles, they have the ability to solubilize oil or other lipophilic substances even when dissolved in polar solvents. In non-polar solvents surfactants form reverse micelles which have the ability to solubilize water or other polar substances when the reaction medium is non-polar. Micellar catalysis is another important area where surfactants find application. Microemulsions which find extensive use in drug delivery, enhanced oil recovery and synthesis of nanoparticles are also formed from oil, water, surfactant and a co-surfactant.

Q. 24. Describe briefly the terms emulsilfication, types of emulsions and emulsifier.

Ans.: An emulsion is a colloidal solution of two or more immiscible liquids which is stabilized by an emulsifier. Surfactants constitute an important class of emulsifiers. Emulsions are usually of two types; oil-in-water (O/W) where oil is the dispersed phase (inner phase) and water is the dispersion medium (outer phase) and water-in-oil (W/O) where water is the dispersed phase (inner phase) and oil is the dispersion medium (outer phase). Oil here refers to a lipophilic substance insoluble in water. O/W emulsions are produced by emulsifying agents that are more soluble in water than in the oil phase whereas W/O emulsions are produced by emulsifying agents that are more soluble in oil than in water phase. This is called Bancroft rule. For example, moisturizing lotions, milk are O/W emulsions while cold cream, butter are W/O emulsions.

Q. 25. What is surface excess? What are its units?

Ans.: A surface active solute tends to accumulate on the surface because by doing so the hydrophobic part, which has unfavourable interactions with water, is projected outside while the hydrophilic part remains in water. Surface excess is defined as the excess amount of solute present per unit area in the surface region as compared to the bulk phase. The units of surface excess are mol m^{-2}.

Q. 26. Write Gibb's adsorption equation and explain its significance.

Ans.: Gibb's adsorption equation relates change in surface tension of a solution to the concentration of solute. According to Gibb's adsorption equation,

$$\Gamma_2 = -\frac{1}{RT}\left(\frac{d\gamma}{d \ln c_2}\right)_T = -\frac{c_2}{RT}\left(\frac{d\gamma}{dc_2}\right)_T$$

Γ_2 is the surface excess, defined as the excess amount of solute present per unit area in the surface region as compared to the bulk phase. c_2 is the concentration of solute, $d\gamma$ is the change in surface tension of solvent due to the presence of solute, R and T are the gas constant and temperature, respectively. Surface active substances tend to accumulate on the surface and reduce the surface tension of water. For such substances, surface tension decreases with increase in concentration, the derivative ($d\gamma/d\ln c_2$) is negative and surface excess Γ_2 is positive.

Q. 27. For which type of substances surface concentration is negative?

Ans.: The most common example is the aqueous solutions of strong electrolytes. With increase in the concentration of strong electrolytes such as NaCl and KCl, the surface tension of water increases indicating a negative surface concentration. This is believed to be the result of interionic electrostatic attraction which tends to bring oppositely charged ions close and away from the surface. In such cases, the derivative ($d\gamma/d\ln c_2$) will be positive in the Gibb's adsorption equation, and surface excess Γ_2 is negative.

Q. 28. What is surface pressure?

Ans.: Surface pressure (π) is defined as the difference between surface tension of solvent and solution at a given solute concentration and temperature. Mathematically, $\pi = \gamma_{Solvent} - \gamma_{Solution}$. Surface pressure ($\pi$), also called film pressure, is a two dimensional pressure and has units of Nm^{-1}.

Q. 29. What is Parachor?

Ans.: Parachor is an empirical constant which relates surface tension to molecular volume. It is defined as $P = (\gamma^{1/4}) M/d$ where $\gamma^{1/4}$ is the fourth root of surface tension, M is the molar mass and d is the density. The quantity M/d represents molar volume and has units of cm^3/mol. A comparison of the parachor values of two liquids effectively means comparison of molar volumes under conditions such that their surface tensions are equal. Since parachor is an additive as well as constitutive property, it is used for determination of some atomic and structural parameters of compounds. It can also be employed for the estimation of surface tension of liquids of known structure and density.

Q. 30. How can we determine the parachor values of solid substances?

Ans.: The solid substance is dissolved in an appropriate solvent and the surface tension and density of the solvent and solution are determined. Since parachor is an additive property, the parachor of solution consisting of a mixture of non-interacting components is given by

$$P_{Soln} = x_1 P_1 + x_2 P_2$$

where P represents the parachor and x the mole fraction of the components of solution. Subscripts 1 and 2 refer to the solvent and solute respectively. Experimental determination of parachors of solvent (P_1) and solution (P_{Soln}), parachor of solid solute (P_2) can be determined.

Q. 31. What is the difference between cohesion and adhesion?

Ans.: Cohesion signifies the force of attraction between like molecules whereas adhesion is the force of attraction between unlike molecules. For example, water has high surface tension due to cohesive forces in water molecules. The meniscus of a liquid in a capillary tube is curved due to adhesive forces between liquid and glass surface of the capillary.

13. Liquid state - Refractive index

Q. 1. What is refraction?

Ans.: Refraction is the change in direction of propagation of a wave when it passes from one medium to another. When light passes from a less dense to a more dense medium (for example passing from air into water), it is refracted (or bent) towards the normal. The normal is a line perpendicular (forming a 90 degree angle) to the boundary between the two substances. The bending occurs because light travels more slowly in the denser medium.

Q. 2. Define refractive index.

Ans.: Refractive index or Index of refraction (n) is given by the ratio of the velocity of light in vacuum to the velocity of light in a specified medium. Refractive index is also defined in another way. If i is the angle of incidence of a ray of light (angle between the incoming ray and the normal) in vacuum and r is the angle of refraction (angle between the ray in the medium and the normal), the refractive index n is defined as the ratio of the sine of the angle of incidence to the sine of the angle of refraction; i.e., $n = \sin i/\sin r$. This is also called law of refraction or Snell's law.

Q. 3. What is the significance of refractive index value?

Ans.: Refractive index of a material is a dimensionless number that describes how light propagates through a medium. The refractive index determines how light is bent or refracted when entering a material. The higher the refractive index of material, the slower the light travels in it and there is correspondingly increased change in the direction of light within the material. For example, the refractive index of water is 1.333, meaning that light travels 1.333 times slower in water than it does in vacuum. Since refractive index is a fundamental physical property of a substance, it is often used to identify a substance, confirm its purity and measure its concentration. Moreover, it is an important property of the components of any optical instrument. It determines the focusing power of lenses and dispersive power of prisms. High refractive index materials are also responsible for lighter and thinner lenses in spectacles and mobile phones.

Q. 4. What are the factors on which refractive index depends?

Ans.: The refractive index of a liquid shows a small decrease with increase in temperature. This is because refractive index usually increases with increase in density of medium and the density of a liquid generally decreases with increase in temperature. The angle of refraction depends on the light speed, which in turn is a function of wavelength and therefore, refractive index value also depends on the wavelength of incident light. The shorter wavelengths are normally refracted more than the longer ones.

Q. 5. Refractive index is sometimes written as n_D. What does D stand for?

Ans.: The subscript D in n_D stands for the D-line of sodium vapor lamp. Sodium emission spectrum exhibits a strong doublet at 589.0 and 589.6 nm. The line at 589.0 nm is most intense. Refractive index measured at a wavelength of 589 nm is usually written as n_D.

Q. 6. How temperature and wavelength are specified in the symbol for refractive index?

Ans.: Refractive index is commonly represented by the symbol n. Generally, temperature and wavelength are also specified in the symbol. The wavelength is written as subscript and the temperature is written as superscript. For example, n^{20}_D signifies that the refractive index values are measured at wavelength 589 nm, the D-line of sodium vapor lamp at a temperature of $20°C$.

Q. 7. How temperature correction can be applied to the measured refractive index value?

Ans.: Most of the refractive index measurements reported in the literature are at 20 or 25⁰ C. The measured refractive index value can be corrected for temperature using the following relationship
$$n_D^{20} = n_D^T + 0.00045\ (T - 20°C)\ or\ n_D^{25} = n_D^T + 0.00045\ (T - 25°C)$$
Here n^T_D refers to the experimental refractive index value measured at temperature T and n^{20}_D and n^{25}_D are the corresponding values at 20 and 25⁰ C, respectively.

Q. 8. Which is the most common type of refractometer used in lab for determination of refractive index of liquids?

Ans.: Various types of refractometers such as handheld/portable manual/digital refractometers, automatic refractometers and general laboratory refractometers are available. However, Abbe's refractometer, named after its inventor Ernst Abbe, is the most common bench top instrument for laboratory measurement of refractive index of a liquid. The Abbe refractometer owes its popularity to convenience in usage, wide range of measurement (n_D = 1.3 to 1.7) and minimal sample requirement. The accuracy of the instrument is about ± 0.0002.

Q. 9. What is the working principle of Abbe's refractometer?

Ans.: Abbe's refractometer working principle is based on the critical angle. The critical angle is the angle of refraction in a medium when the incident radiation approaches the surface of the denser medium at an angle of 90⁰ to the normal that is parallel to the phase boundary. The liquid sample is sandwiched into a thin layer between the two prisms. The lower prism has a rough surface and is the illuminating prism while the upper prism with a smooth surface is the refracting prism. When light passes through illuminating prism the rough surface of the prism acts as the source for an infinite number of rays that pass through the sample at all angels including those almost parallel to the surface. The radiation is then refracted at the interface of the sample and the smooth surface of the refracting (upper) prism. After this it passes into the fixed telescope. The divergent critical angle rays of different colors are collected into a single white beam that corresponds to the D-line of sodium with the help of two Amici prisms that can be rotated with respect to one another. The prism angle is changed until the boundary between the light and dark region just coincides with the cross-wires provided in the eyepiece of the telescope. Samples with different refractive indexes will produce different angles of refraction and this will be reflected in a change in the position of the borderline between the light and dark regions. By appropriately calibrating the scale, the position of the borderline can be used to determine the refractive index of any sample. Measurements at constant temperature can be accomplished by circulation of water through the jacket surrounding the prism. Although Abbe refractometers come in many variants that differ in details of their construction and optical design, nearly all of them utilize this principle. In original design whole telescope was rotated around stationary sample and scale. In modern designs telescope position is fixed, what moves is an additional mirror between sample and telescope.

Q. 10. The speed of light and hence the index of refraction varies significantly with the wavelength of incident light. How the effect of wavelength is taken care of in Abbe's refractometer?

Ans.: The refractive index of a substance is a function of wavelength of light. If the light source is not monochromatic, light gets dispersed and shadow boundary is not well defined. Thus instead of seeing sharp edge between light and dark regions, we see a blurred blue or red border. This effect is referred to as dispersion. The shorter wavelengths are normally refracted more than the longer ones. To prevent dispersion even when using white light, a set of compensating prisms, called amici prisms are introduced into the optical path after the refracting prism in Abbe refractometers. These compensating prisms are designed so that they can be adjusted to correct or compensate for the dispersion of the sample in such a way that they reproduce the refractive index that would be obtained with monochromatic light of wavelength 589 nm, the sodium D line.

Q. 11. What is the highest value of refractive index which can be measured with Abbe refractometer?

Ans.: The refracting prism in Abbe refractometer is made of a glass with high refractive index (e.g., 1.75) and the refractometer is designed to be used for samples having a refractive index smaller than that of the refracting prism.

Q. 12. What precautions should be observed while measuring refractive index using Abbe refractmeter?

Ans.: Following precautions should be observed during refractive index measurement using Abbe refractometer.
i) The prism glass should not be scratched during cleaning or applying sample. ii) The amount of sample should be enough to form a thin film across the whole prism. Special care should be taken when using volatile samples since they evaporate easily. iii) Close to the refractive index of the sample, the hand wheel on the right side of instrument should be adjusted until the dark and light regions are clearly visible. Before taking the final reading, it is also necessary to sharpen the borderline between the light and the dark regions using the compensator dial on the front of the refractometer. For proper focusing, the eye piece which is used to observe the boundary, can also be adjusted by moving it up and down. iv) The refractive index of water at different temperatures is easily available in the literature. The instrument should be calibrated using the known standard refractive index of water at a specified temperature as the reference. All other observed values can then be modified by applying the correction, which is the difference between the actual and the observed value for water.

Q. 13. What are the typical values of refractive index?

Ans.: Refractive index is unity for vacuum. Most transparent media have refractive indices between 1 and 2. Gases at atmospheric pressure have refractive indices close to 1 because of their low density. Almost all solids and liquids have refractive indices above 1.3.

Q. 14. How do you define specific refractivity and molar refractivity mathematically?

Ans.: The widely used equations of Lorentz and Lorentz are used. Specific refractivity is defined as $R_s = 1/d \, \{(n^2 - 1)/(n^2 + 2)\}$, where n is the refractive index and d is the density of the medium. Molar refractivity is defined as $R_m = M/d \, \{(n^2 - 1)/(n^2 + 2)\}$, where M is the molar mass of the substance. The molar refractivity is an additive as well as constitutive property.

Q. 15. Explain temperature-dependence of specific refractivity and molar refractivity.

Ans.: Refractive index, n is temperature-dependent while specific refractivity and molar refractivity are independent of temperature. This is because temperature-dependence of refractive index is primarily due to the temperature-dependence of density of the medium and specific/molar refractivity is defined in such a way that the two effects cancel each other.

Q. 16. How molar refractivity is useful in elucidating the structure of liquids?

Ans.: Molar refractivity is a characteristic of the substance and is temperature-independent. It is an additive and constitutive property. The molar refraction of a substance is the sum of the contributions of atoms (atomic refractions) and bonds (bond refractions) present. On comparing the calculated and experimentally observed values, the structure of the compound can be elucidated.

Q. 17. Explain how molar refractivity of a solid can be determined?

Ans.: To determine molar refractivity of a solid, it is dissolved in a suitable solvent and the refractive index, n and density, d of the solution are determined. The molar refractivity of the solution is then given by the formula

$$R_m \text{ (Solution)} = \left(\frac{n^2 - 1}{n^2 + 2}\right)\left(\frac{x_1 M_1 + x_2 M_2}{d}\right)$$

where x_1 and x_2 are the mole fractions of the solvent and solute, respectively and M_1 and M_2 are their molar masses. Since molar refractivity is an additive property, the total molar refractivity is the sum of the contributions made by solute (solid) and the solvent and therefore,

$$R_m \text{ (Solution)} = x_1 R_1 + x_2 R_2$$

where R_1 is the molar refractivity of the solvent and R_2 that of the solid. Thus knowing R_1 and the experimentally determined value of R_m (Solution), R_2 the molar refractivity of the solid solute can be calculated.

Q. 18. How molar refractivity of individual liquids in liquid mixture can be determined?

Ans.: Since molar refractivity is an additive property, the molar refractivity of a mixture of liquids 1 and 2 is related to the molar refractivities of the individual liquids according to the following equation

$$R_m = x_1 R_m(1) + x_2 R_m(2) \quad \text{(i)}$$

where R_m, $R_m(1)$ and $R_m(2)$ are the molar refractivities of mixture, liquid (1) and liquid (2), respectively and x_1 and x_2 are mole fractions of liquids (1) and (2), respectively. It follows from equation (i) that

$$R_m = x_1 R_m(1) + (1 - x_1) R_m(2) = R_m(2) + x_1 \{R_m(1) - R_m(2)\} \quad \text{(ii)}$$

Since $R_m(1)$ and $R_m(2)$ are constants, a plot of R_m against x_1 will be a straight line. $R_m(1)$ and $R_m(2)$ can be obtained from the slope and intercept of the straight line. Such a straight line plot can also be used as a calibration plot to determine the composition of binary liquid mixture.

Q. 19. Give some examples to support the fact that molar refractivity is an additive and constitutive property.

Ans.: The molar refractivity is known to be additive as well as constitutive. It has been found that the difference in the molar refractivity of the successive members of homologous series is nearly constant. It has also been found that the isomeric compounds with similar structures such as isopropyl alcohol and n-propyl alcohol have nearly same value of molar refractivity. These observations support the additive nature of molar refractivity. Similarly in the case of compounds containing double bond, triple bond or a closed ring, the observed value of molar refractivity is higher than the value obtained by simply adding the atomic refractivities. Thus molar refractivity is also constitutive in nature.

Q. 20. What are the units of refractive index, specific refractivity and molar refractivity?

Ans.: Refractive index is a dimensionless number. It has no units. Specific refractivity has units which are reciprocal of density that is cm^3/g in cgs system and m^3/kg in SI system. Units of molar refractivity are cm^3/mol in cgs system and m^3/mol in SI system.

Q. 21. What are the limitations of Abbe's refractometer?

Ans.: Abbe refractometer has the following limitations. i) Refractive indices of only such substances can be measured whose values are less than the refractive index of the glass of refracting (upper) prism. ii) Because of the energy loss due to the absorption of incident light by the sample, it is difficult to measure the refractive index of dark and absorbing samples. iii) Measured refractive index value is for the liquid layer next to the glass surface. It may be different from that of the bulk of the liquid due to distortion of molecules caused by interaction forces between the glass and the liquid.

Q. 22. What are the relative advantages of digital refractometers over manual analog instruments?

Ans.: Most manual Abbe refractometers are without temperature control and involve difficult sample handling and cleaning. The accuracy of these instruments is also limited to 3 digits. In modern digital refractometers the sample handling is convenient; they have a compact design with temperature-controlled cell. Moreover, compared to an analog instrument, a digital refractometer gives fast results in different measurement units and the data can be easily exported to computer for analysis. The measurement range depends on the specified accuracy; the better the accuracy the narrower is the measurement range.

Q. 23. What is Brix and how Brix values can be measured refractometrically?

Ans.: Degrees Brix (^{0}Bx) is the sugar content of an aqueous solution. One degree Brix is 1g of sucrose in 100g of solution and represents the strength of the solution as percentage by mass. Brix scale is used extensively in the agriculture, food and beverage industries. Since refractive index varies with the concentration of sugar solution, refractometry is the most common method of measuring Brix values. Most refractometers have a dual scale which enables measurement of both refractive index and sugar concentration. The sugar scale can be calibrated to get Brix values.

Q. 24. Does refractive index depend on the angle of incidence?

Ans.: Refractive index is a property of the medium through which light wave passes. It is completely independent of the angle of incidence. As the angle of incidence increases, the angle of refraction also increases proportionally in accordance with the Snell's law. Different wavelengths of light have different refractive indices for the same material.

14. Liquid state - Viscosity

Q. 1. What is fluidity? How fluidity is related to viscosity?

Ans.: Fluidity is defined as the ease with which a liquid flows. It is the reciprocal of viscosity.

Q. 2. What is viscosity?

Ans.: The property that characterizes a fluid's resistance to flow is called its viscosity. It is the reciprocal of fluidity.

Q. 3. Distinguish between laminar and turbulent flow of a liquid.

Ans.: When a fluid is flowing through a closed channel such as a glass tube or between two flat plates, the flow can be either laminar or **turbulent** depending on the velocity with which the fluid is flowing and the viscosity of the fluid. Laminar or streamline flow occurs when a fluid flows in parallel layers, without any lateral mixing between the layers and the motion of the particles of the fluid is very orderly. Laminar flow tends to occur at lower velocities of the fluid. Above a threshold velocity it becomes turbulent. Turbulent flow is a less orderly flow and is characterized by lateral mixing of the layers. In non-scientific terms, laminar flow is smooth while turbulent flow is rough.

Q. 4. Define coefficient of viscosity.

Ans.: According to Newton's law of viscosity, the frictional force exerted by slower moving fluid layer on faster moving layer, F_z per unit area A of fluid layer is directly proportional to the velocity gradient dv_x/dz. Velocity gradient is the difference in velocity of adjacent layers divided by the distance between layers. It may be noted that the frictional force is perpendicular to the direction of flow of fluid. If fluid is flowing in the x-direction, the frictional force acts in the z-direction.

$$F_z \, \alpha \, A \, dv_x/dz \text{ and } F_z = - \eta \, A \, dv_x/dz$$

Negative sign indicates that the frictional force opposes the velocity gradient. The proportionality constant, η is called the coefficient of viscosity of the fluid. It can be defined as force per unit area required to maintain unit velocity gradient between adjacent layers of fluid.

Q. 5. What are Newtonian and non-Newtonian fluids?

Ans.: Liquids which obey Newton's law of viscosity are called Newtonian fluids. A non-Newtonian fluid is a fluid that does not follow Newton's law of viscosity. In Newtonian liquids, the coefficient of viscosity is independent of the velocity gradient while in non-Newtonian liquids, the coefficient of viscosity decreases with increase in the velocity gradient. All gases and most liquids which have simple molecular formula and low molecular weight such as water, benzene, ethyl alcohol, hexane and most solutions of simple relatively low molecular weight substances are Newtonian fluids. Generally non-Newtonian fluids are complex high molecular weight substances such as polymer solutions and mixtures such as slurries, pastes, gels etc.

Q. 6. What are cgs and SI units of viscosity?

Ans.: The cgs unit for viscosity is poise (P) which is named after Jean Poiseuille. It is more commonly expressed as centipoise (1 cP = 0.01 poise). The SI unit of viscosity is Pascal-second (Pa s). 1 P = 0.1 Pa s = 0.1 kg m^{-1} s^{-1}. 1 cP = 1 mPa s (one millipascal second).

Q. 7. What is shear, shear stress and shear gradient or shear rate? If shear stress at a point in a liquid is 0.03 N/m² and the velocity gradient at the point is 0.15 s⁻¹, what is the coefficient of viscosity in poise?

Ans.: The frictional resistance between adjacent layers of liquid flowing through a cylindrical pipe generates a velocity gradient. The kind of deformation of a liquid produced by velocity gradient is called shear. Shear stress is defined as the viscous or frictional force per unit area. The velocity gradient is also called shear gradient or shear rate. Accordingly the Newton's law of viscosity can also be stated as 'the shear stress between adjacent fluid layers is proportional to the negative value of the shear gradient between the two layers'.

$$\frac{F_z}{A} = -\eta \left(\frac{dv_x}{dz}\right), \eta = -\frac{F_z/A}{dv_x/dz}$$

Shear stress
$F_z/A = 0.03 N/m^2, dv_x/dz = -0.15\ s^{-1}$

(Negative sign is due to the fact that the viscous force hinders the flow of the fluid).
Coefficient of viscosity
$\eta = 0.03/0.15 = 0.2\ Ns/m^2 = 0.2\ Pa\ s = 2\ poise$

Q. 8. What are the various factors on which the viscosity of a liquid depends?

Ans.: Viscosity of a liquid depends on the nature of the liquid. For example the size and shape of its molecules, molecular structure and strength of intermolecular forces involved. Liquids whose molecules are polar and have the ability to form inter- and intramolecular hydrogen bonds such as glycerol are more viscous than the non-polar molecules of the same size. Similarly liquids containing long molecules such as fuel oils are more viscous because the molecular chains get tangled up in each other. Amongst external factors, viscosity of a liquid depends strongly on temperature, generally becoming smaller as temperature is increased. This is because increased kinetic motion at higher temperatures decreases the intermolecular interactions between adjacent layers of fluid. The viscosity of gases depends both on temperature and pressure.

Q. 9. Which empirical equation describes the temperature-dependence of viscosity of liquids?

Ans.: Viscosity decreases considerably with rise in temperature (roughly about 2% per degree) and follows fairly closely the Arrhenius-like equation called Andrade equation according to which

$$\eta = A\, e^{B/RT}$$

where A and B are constants for a given liquid. This equation is nearly identical to the Arrhenius equation that describes the temperature variation of rate constant (k) of a chemical reaction, except that it does not have a negative sign in the exponential term. By analogy with the Arrhenius theory of reaction rates, B which has dimensions of work can be regarded as the activation energy for viscous flow.

Q. 10. Do gases have viscosity? What is the difference between liquid viscosity and gas viscosity?

Ans.: Yes. Gases are also fluids and all fluids experience viscous force. However, in liquids since molecules are close to each other, intermolecular attractive forces are mainly responsible for viscosity whereas in gases since molecules are far apart, intermolecular forces are negligible and random intermolecular collisions between the gas molecules are mainly responsible for viscosity. Due to the absence of intermolecular attractive forces, gases are less viscous than liquids. Most ordinary liquids have viscosities of the order of 1 to 1000 mPa s while for gases it is of the order of 1 to 10 μPa s. This difference also reflects in the temperature-dependence of viscosity. While the viscosity of a liquid decreases with increase in temperature, that of a gas increases with increase in temperature. This is because the cohesive intermolecular forces decrease with increase in temperature in liquids while collision frequency increases with increase in temperature in gases.

Q. 11. Why glycerol is more viscous than water?

Ans.: The answer lies in the difference in structure of the two liquids. Glycerol has three hydroxyl groups per molecule compared to one in water. More intermolecular and intra-molecular hydrogen bonding interactions make glycerol more viscous.

Q. 12. Define Poiseuille's law of viscosity?

Ans.: The rate of flow of a liquid through a pipe depends on the following factors: the pressure head (P), the length of the pipe (l), radius of the pipe (r) and the coefficient of viscosity of the liquid (η). According to Poiseuille's law of viscosity, for laminar flow of a Newtonian liquid through a pipe, the rate of flow (V/t), i.e., the volume (V) of liquid flowing per unit time (t) is given by $V/t = \pi P r^4/8\eta l$. Non-Newtonian liquids do not obey Poiseuille's law because their viscosities are not independent of the velocity gradient.

Q. 13. Describe the most commonly used method for experimental determination of coefficient of viscosity of a liquid.

Ans.: Capillary viscometers are the most frequently used instruments for laboratory determination of the coefficient of viscosity since they are easy to use and comparatively inexpensive. The principle of measurement is based on the Poiseuille's law of viscosity. It involves the determination of time required for a given volume of liquid to flow through a capillary tube of uniform cross section. Many capillary tube viscometers have been devised but Ostwald and Ubbelohde viscometers are amongst the most frequently used. Ostwald viscometer is a U-shaped piece of glassware with a reservoir on one side and a measuring bulb with a capillary on the other. A liquid introduced into the reservoir is sucked through the capillary into the measuring bulb. The liquid is allowed to travel back through the measuring bulb and the time taken by the liquid to pass through two calibrated marks is used to measure the coefficient of viscosity. The Ubbelohde viscometer is closely related to the Ostwald viscometer. The Ubbelohde device has a third arm extending from the end of the capillary and open to the atmosphere. In this way the pressure head only depends on a fixed height and no longer on the total volume of liquid. This is usually used as a relative method. The rate of flow of test liquid and a reference liquid are measured. According to Poiseuille's law, $V/t = \pi P r^4 / 8\eta l$ or $\eta = \pi P r^4 t / 8Vl$.

The pressure head, P = hdg where h, d, g are the height of the liquid column, density of the liquid and the acceleration due to gravity, respectively. If η_0 is the coefficient of viscosity of the reference liquid, on taking the ratio η/η_0, many terms are cancelled out and we get the simple relationship,

$\eta/\eta_0 = t/t_0 \times d/d_0$

Thus η can be determined using known value of η_0 for the reference liquid. For aqueous solutions and most other liquids, the reference liquid is taken as water for which the coefficient of viscosity and density at different temperatures are easily available in the literature.

Q. 14. Same volume of an organic liquid and water take 70 and 100 seconds, respectively to flow through an Ostwald viscometer at the same temperature. If densities of organic liquid and water are 0.80 and 1.00 g/cm³ and viscosity of water is 10.0 cP, calculate the viscosity of the organic liquid in SI units.

Ans.: The ratio of viscosities of organic liquid and water is given by

$$\frac{\eta_{org}}{\eta_w} = \frac{t_{org}}{t_w} \times \frac{d_{org}}{d_w} \text{ or or } \eta_{org} = \eta_w \times \left(\frac{t_{org}}{t_w} \times \frac{d_{org}}{d_w}\right) = 10 \times \frac{70}{100} \times \frac{0.80}{1.00} = 5.6 \, cP$$

1 P = 0.1 Pa s and 1 cP = 1 mPa s

η_{org} = 5.6 mPa s = 5.6 × 10⁻³ Pa s.

Q. 15. What precautions should be observed for the determination of coefficient of viscosity using Ostwald viscometer?

Ans.: 1) The viscometer should be clamped vertical, 2) Same volume of liquid should be used for the reference and test liquid so that the pressure head remains same, 3) Temperature should be kept constant by keeping the viscometer in a thermostat, 4) Liquid should be filtered before measurement to get rid of any suspended impurities which may hinder the flow, 5) The radius of the capillary should be small to ensure laminar flow since Poiseuille's law is not applicable for turbulent flow.

Q. 16. What is Stoke's law?

Ans.: Stoke's law is a well-established hydrodynamic law according to which a spherical solid falling in a fluid experiences a frictional force, $F_{fr} = fv = 6\pi\eta rv$ where f is the frictional coefficient, v is the velocity with which the solid is falling, η is the coefficient of viscosity of the fluid and r is the radius of the sphere.

Q. 17. Describe method based on Stoke's law for experimental determination of viscosity of liquids.

Ans.: Using Stoke's law it can be shown that the rate of fall (v) of a sphere (radius r, density ρ_1) through a liquid (density ρ_2) is given by

$$v = \frac{2(\rho_1 - \rho_2) r^2 g}{9 \eta}$$

The symbols g and η represent the acceleration due to gravity and coefficient of viscosity of liquid, respectively. Here again relative determinations with the same tube and ball largely simplify the method. The method is particularly suitable for viscous oils available in large quantity. For any oil lubrication system, the oil viscosity is the most important parameter.

Q. 18. What are various ways by which viscosity can be expressed? Also write the units in each case.

Ans.: 1) Coefficient of viscosity (η). Units: The SI unit of viscosity is Pa s. Other commonly used units are given in Q.6

2) Relative viscosity (η_{rel}) = η/η_0 where η_0 is the coefficient of viscosity of reference liquid, usually water. Units: Relative viscosity has no units since it is a ratio.

3) Specific viscosity (η_{sp}) = $\eta_{rel} - 1$. Units: Specific viscosity also has no units.

4) Reduced viscosity (η_{red}) = η_{sp}/c where c is the concentration of solution. Units: Inverse of concentration, L mol^{-1}.

5) Intrinsic viscosity $([\eta_{int}]) = \lim_{c \to 0} \frac{\eta_{sp}/c}$. The y-intercept of a plot of η_{sp}/c versus c is equal to the intrinsic viscosity. Units: L mol^{-1}.

Q. 19. What is the difference between dynamic viscosity and kinematic viscosity?

Ans.: Dynamic viscosity is a measure of the fluid's resistance to flow when an external force is applied while kinematic viscosity is a measure of the fluid's resistance to flow under the weight of gravity. Thus kinematic viscosity measures the fluid's inherent resistance to flow when no external force, except gravity, is acting on it. Two fluids that have the same dynamic viscosity can have different kinematic viscosities and vice versa. This is because dynamic viscosity is independent of density while kinematic viscosity is dependent on the density of the fluid.

Q. 20. What are the units of kinematic and dynamic viscosity? How can kinematic viscosity be converted to dynamic viscosity and vice versa?

Ans.: The cgs units of kinematic and dynamic viscosity are centistokes (CSt) and centipoise (cP), respectively while the corresponding SI units are m^2/s and Pa s. The formula for the inter conversion of the two types of viscosities is

Kinematic viscosity x Density = Dynamic viscosity

Q. 21. For what type of liquids dynamic viscosity should be measured?

Ans.: The measurement of dynamic viscosity is most useful for non-Newtonian liquids which change their apparent characteristics when force or pressure is applied. Kinematic viscosity, on the other hand, is used mostly for Newtonian liquids for which there is no change in viscosity as the applied force is changed.

Q. 22. Distinguish between monodisperse and polydisperse polymers.

Ans.: A polymer sample in which all the molecules have same molecular weight is said to be a monodisperse polymer whereas a polydisperse polymer has molecules of different molecular weights. Nearly all natural polymers such as proteins are monodisperse while most synthetic polymers are polydisperse.

Q. 23. What do you understand by the number-average and weight-average molecular weight of a polymer?

Ans.: The molecular weight of a synthetic polymer does not have a single value since all molecules do not have same chain length and the extent of branching of chains. The distribution of molecular weight in a polymer sample is defined in terms of some kind of average; the two most common being the number-average molecular weight and weight-average molecular weight. The number average molecular weight is the total weight of the sample divided by the number of molecules in the sample. The total weight of the sample is the summation of the product of the weight of the molecule and the number of molecules of that weight. If there are N_1 molecules of weight M_1, N_2 molecules of weight M_2, N_3 molecules of weight M_3 and so on, the total weight of the sample $= \Sigma N_i M_i$ and the total number of molecules $= \Sigma N_i$. Number-average molecular weight, $M_n = \Sigma N_i M_i / \Sigma N_i$.

The weight-average molecular weight is based on the fact that a bigger molecule contains more of the total mass of the polymer sample than the smaller molecules. In other words, weight-average molecular weight depends not only on the number of molecules present but also the weight of each type of molecule. Thus N_i is replaced by $N_i M_i$ in the expression for the number-average molecular weight and weight-average molecular weight, $M_w = \Sigma N_i M_i^2 / \Sigma N_i M_i$. The term $N_i M_i / \Sigma N_i M_i$ represents the weight fraction (w_i), the fraction of the total weight represented by each type of molecule. The weight-average molecular is then calculated by taking the summation, $\Sigma w_i M_i$.

Q. 24. What do you understand by the molecular weight distribution of a polymer?

Ans.: In synthetic polymers, all molecules do not have same degree of polymerization and therefore, there is a distribution of molecular weights around an average value. It is usually represented as a bell shaped plot of the number of molecules of a given molecular weight against molecular weight.

Q. 25. What is polydispersity index and what does it represent?

Ans.: Polydispersity index is a measure of the distribution of molecular mass in a given polymer sample. It is calculated as M_w/M_n, the ratio of weight-average molecular weight and number-average molecular weight. Polydispersity index is unity for a monodisperse sample and greater than one for a polydisperse sample.

Q. 26. Define viscosity-average molecular weight and explain how it can be determined?

Ans.: Molecular weight can also be calculated from the viscosity of a polymer solution. It is based on the principle that the contribution of bigger polymer molecules to the viscosity of solution is more than that of small ones. The molecular weight obtained by viscosity measurement is closer to the weight average than the number average molecular weight but is different from either of these.

Viscosity-average molecular weight can be determined using the empirical equation called Mark-Houwink equation according to which intrinsic viscosity, $[\eta_{int}] = KM^a$ where K and a are constants for particular polymer-solvent system and M is the molar mass of the polymer. For a large number of systems, 'K' and 'a' values are available in the tables.

Q. 27. Why different liquids have different viscosities under same conditions of temperature and pressure? What information about the liquid can be obtained from viscosity measurements?

Ans.: Under identical conditions, molecular structure which determines the extent of intermolecular interactions as well as size and shape of the liquid molecules determines the viscosity of a liquid. Molecules of smaller size with roughly spherical shape and lower ability for intermolecular interactions are less viscous.

Q. 28. How can we determine the molecular diameter of a polymer from viscosity measurements?

Ans.: The method is based on the assumption that the polymer molecule is coiled in roughly spherical shape. According to the Einstein equation, for large rigid spheres, the relative viscosity is given by $\eta_{rel} = 1 + 2.5\,\varphi$, where φ is the volume fraction of solute (the fraction of the total volume occupied by solute particles) in solution. The value of φ may be calculated from the concentration (c) of solute particles in g/mL and the partial specific volume (v) of the solute. The volume fraction, $\varphi = cv$. Partial specific volume v is defined as the volume increment resulting from the addition of 1g of solute to a very large volume of solution. From Einstein equation,

$$\eta_{rel} - 1 = \eta_{sp} = 2.5\varphi = 2.5\,\bar{c}\bar{v}.\quad \frac{\eta_{sp}}{c} = \eta_{red} = 2.5\,\bar{v}.\ \ \text{limit}_{c \to 0}\ \frac{\eta_{sp}/c}{} = [\eta_{int}] = 2.5\,\bar{v}.$$

For a spherical particle, the volume of 1 molecule = $4/3\,\pi r^3$, volume of 1 mol of solute = $4/3\,\pi r^3 N$, where N is the Avogadro's number. Volume of 1g of solute = $4/3\,\pi r^3 N/M_n$ where M_n is the number-average molecular weight. Relative viscosities of the polymer solutions of different concentrations are determined and η_{sp}/c values are plotted against the concentration of solution. Straight line so obtained when extrapolated to zero concentration gives intrinsic viscosity. Knowing the intrinsic viscosity and the number-average molecular weight of the polymer, average radius of polymer molecules in solution can be determined. It may be mentioned that the radius or diameter of polymer obtained by this method is only approximate since the polymer molecules are assumed to be rigid spheres.

Q. 29. What do you understand by excluded volume?

Ans.: Excluded volume of a molecule is the volume that is inaccessible to the other molecules in the system as a result of the presence of first molecule. For a pair of hard spheres the excluded volume is four times the volume of the spheres. Excluded volume concept is particularly relevant in polymer systems. A polymer chain cannot coil altogether randomly because it is subjected to the restriction that no two parts of the chain may be at the same point in space at the same time. This restriction becomes more important for a high molecular weight polymer in a good solvent.

Q. 30. Give some applications of viscosity measurements.

Ans.: Viscosity measurements have many applications in diverse fields. Some of these are given here. i) Determination of molecular weight of polymeric and some other liquids, ii) The knowledge of the coefficient of viscosity and its variation with temperature helps us choose a suitable lubricant for specific machines, iii) Determination of flow characteristics of liquids are important in various industrial processes such as paints and varnishes, cough syrups, petroleum products, cosmetics etc.

15. Solutions – Partial molar volume

Q. 1. What are partial molar quantities?

Ans.: In thermodynamics, an open system is one which can exchange both mass and energy with the surroundings. Therefore, extensive properties in open systems are function of not only temperature and pressure but also the composition of the system that is the number of moles of various components in the system. The change in any extensive property of a large system of definite composition accompanying the addition of one mole of a component at constant temperature and pressure is called the partial molar property of the component. Partial molar properties are represented by writing a bar over the symbol of the property. For example, partial molar value for component i of any property X can be written as

$$\bar{X}_i = \left(\frac{\partial X}{\partial n_i}\right)_{T, P, n_j}$$

where n is the number of moles and subscript j stands for the number of moles of all other components except component i.

Q. 2. How partial molar value of a property is related to the total value of the property?

Ans.: Partial molar property is the contribution per mole of each particular constituent of the mixture to the total value of the property. The total value of property
$$X = \sum \bar{X}_i n_i$$
For example, total volume, $V = \sum \bar{V}_i n_i$

Q. 3. Define partial molar volume and explain its importance.

Ans.: Partial molar volume is broadly understood as the contribution of a component of mixture to the overall volume of solution. It may be defined as the increase in the volume when 1 mole of component i is added at constant temperature, pressure and number of moles of other components present to such a large amount of solution that insignificant concentration change results. Mathematically it can be expressed as

$$\bar{V}_i = \left(\frac{\partial V}{\partial n_i}\right)_{T, P, n_j}$$

Partial molar volumes and their deviations from the values expected for ideal solutions are of particular interest in the theory of solutions. The importance of partial molar volumes is also due to their thermodynamic connection with other important quantities such as chemical potential.

Q. 4. With the help of an example explain why the total volume of solution is not always equal to the sum of the volumes of solute and solvent?

Ans.: Let us take the example of an aqueous solution of ethanol. If we mix 100g of ethanol (= 126.7 mL at 20°C) and 900g of water (= 901.6 mL at 20°C) we obtain a solution with density 0.982 g/mL and total volume of 1018.3 mL whereas the additive volume of the two components is 1028.3 mL. The decrease in volume of 10 mL ($\approx$ 1%) on mixing the two components is due to the difference in the solute-solvent and solvent-solvent intermolecular interactions. The packing arrangement of the solute molecules in the solvent is different from the packing of solvent molecules in solvent. Thus the total volume of solution is not equal to the sum of the volumes of solute and solvent and the magnitude and sign of the increase/decrease in volume is an indication of the nature of intermolecular interactions involved. That is why the standard method of preparing solutions of accurately known concentrations is not the addition of a weighed amount of solid to a known volume of solvent, but the addition of solvent with mixing to a weighed amount of solid until the final volume reaches the desired value.

Q. 5. How partial molar volume of sodium chloride in aqueous solution can be determined from density measurements?

Ans.: The total volume of an amount of solution containing 1 kg of water and m moles of solute is given by

$$V = n_1 \bar{V}_1 + n_2 \bar{V}_2 = 55.51 \bar{V}_1 + m \bar{V}_2 \quad (1)$$

Where n_1 and n_2 are the number of moles and $\bar{V}_1$ and $\bar{V}_2$ are the partial molar volumes of the solvent and solute, respectively. The number of moles of solvent in 1 kg of solvent (water), $n_1 = 1000/18 = 55.51$. The number of moles of solute in 1 kg of solvent, $n_2 = m$, the molality of the solute in solution. The partial molar volume of the solute $\bar{V}_2$ can be determined from the apparent molar volume ϕ. The apparent molar volume of the solute (ϕ) is calculated on the assumption that the molar volume of the solvent remains constant (is the same in solution as it is in the pure solvent) and any change in volume is due only to the solute. This assumption is not strictly true but it fits a mathematically simplified picture of solute-solvent interactions. Let $\bar{V}_1^0$ be the molar volume of pure water. Then we define the apparent molar volume, ϕ of the solute by the equation

$$V = n_1 \bar{V}_1^0 + n_2 \phi = 55.51 \bar{V}_1^0 + m\phi \quad (2)$$

which can be rearranged to give

$$\phi = \frac{1}{n_2}(V - n_1 \bar{V}_1^0) = \frac{1}{m}(V - 55.51 \bar{V}_1^0) \quad (3)$$

Now the total volume of solution,

$$V = Total\ mass/Density = \frac{1000 + mM_2}{d} cm \quad (4)$$

and the volume of solvent (water),

$$n_1 \bar{V}_1^0 = 1000/d_0\ cm \quad (5)$$

where d and d_0 are the density of solution and pure water, respectively in units of g cm^{-3} and M_2 is the solute molecular weight in grams. Substituting equations (4) and (5) into equation (3), we obtain

$$\phi = \frac{1}{d}\left(M_2 - \frac{1000(d-d_0)}{m\, d_0}\right) \quad (6)$$

Equation (6) can be written in terms of weights as well since total volume is fixed.

$$\phi = \frac{1}{d}\left(M_2 - \frac{1000(W-W_e) - (W_0 - W_e)}{m(W_0 - W_e)}\right) = \frac{1}{d}\left(M_2 - \frac{1000(W-W_0)}{m(W_0 - W_e)}\right) \quad (7)$$

In eq. (7), the directly measured weights of the pycnometer; W_e when empty, W_0 when filled to the mark with pure water, and W when filled to the mark with the solution are used. Equation (7) is preferable to equation (6) for the calculation of ϕ as it avoids the unnecessary computation of small difference $(d - d_0)$ in densities with high precision.

Now by the definition of partial molar volumes and by use of equations (1) and (2), we can derive the required relations.

$$V = n_1 \bar{V}_1^0 + n_2 \phi$$

$$\bar{V}_2 = \left(\frac{\partial V}{\partial n_2}\right)_{T, P, n_1} = \phi + n_2 \frac{\partial \phi}{\partial n_2} = \phi + m \frac{\partial \phi}{\partial m} \quad (8)$$

We are considering m moles of solute and therefore, $n_2 = m$ in equation (8).

$$V = n_1 \bar{V}_1 + n_2 \bar{V}_2$$

$$\bar{V}_1 = \frac{1}{n_1}(V - n_2 \bar{V}_2) = \frac{1}{n_1}(n_1 \bar{V}_1^0 + n_2 \phi - n_2 \bar{V}_2)$$

Substituting for $\bar{V}_2$ from eq. 8,

$$\bar{V}_1 = \frac{1}{n_1}\left(n_1 \bar{V}_1^0 - n_2^2 \frac{\partial \phi}{\partial n_2}\right) = \bar{V}_1^0 - \frac{m^2 \, d\phi}{55.51 \, dm} \quad (9)$$

We may proceed by plotting ϕ versus m, but the plot is not linear and tangents have to be drawn at desired concentrations to measure slope. However, for solutions of simple uni-univalent electrolytes it has been found that ϕ varies linearly with the square root of molal concentration, $\sqrt{m}$ even up to moderate concentrations. This behavior is in agreement with the prediction of the Debye-Huckel theory for dilute solutions. It can be shown that*

$$\bar{V}_2 = \phi^0 + \frac{3\sqrt{m}}{2} \frac{d\phi}{d\sqrt{m}} \quad (11)$$

$$\bar{V}_1 = \bar{V}_1^0 - \frac{m}{55.51}\left(\frac{\sqrt{m}}{2} \frac{d\phi}{d\sqrt{m}}\right) \quad (12)$$

where ϕ^0 is the apparent molar volume extrapolated to zero concentration. Now we can plot ϕ versus $\sqrt{m}$ and draw the best straight line through the points. For aqueous solutions at 25°C, $\bar{V}_1^0$ the molar volume of pure water is 18 mL/mol. Thus from the slope $d\phi/d\sqrt{m}$ and the ϕ^0 value obtained from intercept, both $\bar{V}_1$ and $\bar{V}_2$ can be determined.

* Refer to Shoemaker D. P., Garland C. W., Nibler J. W. Experiments in physical chemistry. 5th Edition, McGraw-Hill Book Company, Singapore, 1989, pp187-191.

Q. 6. What is the difference between molar volume, partial molar volume and apparent molar volume?

Ans.: The main difference between molar volume and partial molar volume is that molar volume of a substance is the volume of one mole of that substance whereas partial molar volume is the change in volume of a mixture when one mole of a component is added to that mixture at constant temperature and pressure. Partial molar volume is a thermodynamic quantity which describes the non-ideal behavior of the system. The total volume deviates from the sum of the individual volumes due to intermolecular interactions. The apparent molar volume of the sample is calculated on the assumption that the molar volume of the solvent in pure solvent is same as that of the solvent in solution and any change in volume is due only to the sample. This assumption is only approximately true in dilute solutions where a simplified model of solute-solvent interactions assumes change in solute properties only.

Q. 7. Are partial molar volumes always positive?

Ans.: Volume is always a positive quantity but partial molar volume can be positive or negative. For example the partial molar volume of magnesium sulphate at infinite dilution is minus 14 cm^3 per mole which means that when 1 mole of magnesium sulphate is added to water the overall volume decreases. The decrease in volume is due to the partial breakdown of the structure of water in the presence of ions of the salt because some water molecules come out of the regular tetrahedral network and are oriented around ions. The sign and magnitude of partial molar volume is a good indication of the intermolecular interactions involved. At infinite dilution solute-solvent interactions are predominant, solute-solute interactions are almost eliminated since solute molecules are far apart.

Q. 8. Can we measure partial molar volumes of individual ions?

Ans.: Partial molar volumes can be measured only for salts, not individual ions. However, some individual ions are assigned values based on the convention that the partial molar volume of hydrogen ion in aqueous solution is 0 mL/mol. The values for ionic sizes thus obtained are useful for comparative purposes only. Also they are different from those obtained from crystalline radii because the ions are solvated in solution. The partial molar volumes at infinite dilution of most multiply charged cations are negative. The partial molar volumes of most anions at infinite dilution are positive.

Q. 9. What are partial specific properties?

Ans.: Partial specific property is defined as the partial derivative of the property with respect to the mass of the component of interest. For example, partial specific volume,

$$\bar{v}_i = \left(\frac{\sigma V}{\sigma m_i}\right)_{T, P, m_j \neq i}$$

Partial specific quantities have an advantage over partial molar quantities because they can be evaluated without knowledge of the molar mass. For instance, the partial specific volume of a solute is used to determine its molar mass by sedimentation equilibrium method. A partial specific quantity of a substance can also be defined as the partial molar quantity divided by the molar mass. Partial specific volume for example, has dimensions of volume divided by mass and is usually expressed in units of mL/g whereas partial molar volume is expressed as mL/mol.

Q. 10. What is an excess molar quantity?

Ans.: Excess molar quantities are the properties which characterize the non-ideal behaviour of real solutions. They are defined as the difference between the partial molar property of a component in a real solution and the value that would exist in an ideal solution under same conditions.

16. Adsorption

Q. 1. What is the difference between absorption and adsorption?

Ans.: In absorption, the substance is uniformly distributed into the bulk of the solid or liquid. For example, water vapour is absorbed by anhydrous calcium chloride; ammonia is absorbed by water to form ammonium hydroxide. Adsorption, on the other hand, is a surface phenomenon; the adsorbate is retained only on the surface, it does not penetrate into the bulk or interior of the solid or liquid. For example water vapour is adsorbed by silica gel, ammonia is adsorbed by charcoal.

Q. 2. Distinguish between adsorbent and adsorbate in the phenomenon of adsorption.

Ans.: Adsorbent is the solid or liquid on which adsorption takes place while adsorbate is the substance which gets adsorbed.

Q. 3. Why does adsorption occur on the surface of a solid?

Ans.: Adsorption occurs on a surface as a consequence of surface energy. Atoms on the surface experience a bond deficiency because they are not surrounded by other atoms on all sides. In other words, there are unbalanced forces on the surface and thus the surface has a tendency to attract other molecules.

Q. 4. Why adsorption is always an exothermic process?

Ans.: One of the basic principles of physical chemistry is that for a process to be spontaneous, the free energy change should be negative. Since adsorption is a spontaneous process, the free change involved should be negative. A familiar expression in physical chemistry is the relationship

$\Delta G = \Delta H - T\Delta S$ which is the relationship between free energy change (ΔG), enthalpy change (ΔH) and entropy change (ΔS) for a system at equilibrium. Since adsorption is always accompanied by decrease in entropy, ΔS is negative and the entropy factor - $T\Delta S$ is positive. Therefore, ΔH has to be negative for ΔG to be negative. Negative value of ΔH means that the adsorption process must be exothermic, that is adsorption proceeds with the evolution of heat.

Q. 5. What do you understand by adsorption equilibrium?

Ans.: Like a chemical reaction, adsorption is a dynamic process and therefore adsorption equilibrium is attained when the rate of adsorption on the surface is equal to the rate of desorption from the surface.

Q. 6. What are the factors on which adsorption at solid-gas interface depends?

Ans.: Adsorption at solid-gas interface depends on the following factors. i) Nature of the gas and the solid adsorbent, ii) Temperature, iii) Pressure, iv) Porosity or pore size distribution of adsorbent and hence the surface area available for adsorption.

Q. 7. What is activated charcoal?

Ans.: Activated charcoal is a form of processed carbon. The treatment of charcoal, usually with oxygen, generates small micropores on the surface which results in significant increase in the specific surface area available for adsorption and other chemical reactions. Activated charcoal is commonly prepared by two basic processes i) gas activation method and ii) chemical activation method. In gas activation method, charcoal is subjected to oxidizing gases usually carbon dioxide or steam at 800-1000°C. In chemical method, both carbonization and activation processes are carried out simultaneously. Charcoal is mixed with the chemical agent such as zinc chloride, phosphoric acid or alkali metal hydroxide in proper ratio. The mixture is heated at temperatures up to 800°C in the absence of the air (under vacuum or inert gas).

Q. 8. Why activated charcoal is a very good adsorbent?

Ans.: Activated charcoal is a much better adsorbent than simple charcoal because activation creates micropores on the surface resulting in significant increase in the specific surface area of charcoal. Moreover, raw charcoal prepared from the carbonaceous materials such as wood, may contain some impurities in the pores which are also cleared by activation.

Q. 9. Name some good adsorbents other than charcoal.

Ans.: Good adsorbents, other than charcoal include activated silica gel, alumina gel, zeolites, polymers and resins, clays etc.

Q. 10. What is an adsorption isotherm?

Ans.: Adsorption isotherm is a plot of the amount adsorbed versus pressure at constant temperature for adsorption at gas-solid interface and the amount adsorbed versus equilibrium concentration at constant temperature for adsorption at solid-solution interface.

Q. 11. Other than adsorption isotherms, what are the various ways by which adsorption data can be expressed?

Ans.: The other ways by which adsorption data can be expressed are adsorption isobar and adsorption isostere. Adsorption isobar shows the variation of the amount adsorbed with temperature at constant pressure. Adsorption isostere is a curve which shows variation of pressure with temperature, when the amount adsorbed is kept constant. However, adsorption isotherm is the most convenient and frequently used method of expressing adsorption data.

Q. 12. How pores in adsorbents are classified?

Ans.: Pores in adsorbents are classified as micropores, mesopores and macropores according to pore sizes (pore diameters) as given below.
  Micropores < 2 nm, Mesopores 2 – 50 nm, Macropores > 50 nm.

Q. 13. Describe briefly how adsorption isotherms are classified into various types?

Ans.: On the basis of shapes and adsorption characteristics, adsorption isotherms are classified into five types (Type I – V). Type I corresponds to monolayer adsorption, type II to multilayer adsorption, type III also corresponds to multilayer adsorption but with smaller intensity of adsorbate-adsorbent interactions, type IV and V reflect capillary condensation phenomenon.

Q. 14. What is the phenomenon of capillary condensation?

Ans.: Capillary condensation phenomenon is observed during multilayer adsorption of vapour on a porous adsorbent. The pore spaces become filled with condensed liquid from the vapor even below the saturation vapor pressure of the pure liquid. Increased van der Waals interactions between vapor phase molecules inside the confined space of the capillary pores are believed to be responsible for this unique phenomenon.

Q. 15. What is hysteresis effect in adsorption?

Ans.: Type II adsorption isotherms, obtained for adsorption of gases on microporous adsorbents, are reversible that is the adsorption isotherms can be retraced on desorption. However, type IV and V adsorption isotherms obtained for adsorption of gases on mesoporous adsorbents, are not reversible. Such systems show hysteresis effect which means that the adsorption and desorption isotherms follow different paths. The capillary condensation phenomenon associated with these systems is responsible for the appearance of hysteresis effect.

Q. 16. What do you understand by monolayer adsorption?

Ans.: For monolayer adsorption, type I adsorption isotherm is obtained. The amount adsorbed increases with increase in pressure/concentration only up to a certain point beyond which saturation state is reached. At this point the surface is covered by a monolayer of adsorbate. Type I adsorption isotherms are obtained for the adsorption of gases on microporous solids whose pore sizes are not much larger than the molecular diameter of adsorbate and for the adsorption on solid adsorbent from solution phase.

Q. 17. Describe the most common theories proposed for monolayer adsorption?

Ans.: Langmuir and Freundlich adsorption isotherms are the most common theories for monolayer adsorption. Freundlich adsorption isotherm is an empirical relation between amount adsorbed and the equilibrium pressure or concentration. It is mathematically expressed as $x/m = K\, p^{1/n}$ or $x/m = K\, c^{1/n}$. x/m is the amount of solute adsorbed per gram of adsorbent, p is the equilibrium pressure for adsorption at solid-gas interface and c is the concentration of solute in contact with solid surface at equilibrium for adsorption from solution phase. K and n are constants characteristic of the adsorbate-adsorbent system. Magnitude of constant K is an indicator of adsorption capacity while $1/n$ is a measure of the intensity of adsorption. Generally n is < 1 and $1/n$ is > 1. Higher the value of $1/n$, more favourable is adsorption. At high pressure $1/n \approx 0$ and adsorption becomes independent of pressure. The major limitation of Freundlich adsorption isotherm is that it is applicable over a limited range of pressure.

Langmuir adsorption model is based on the following assumptions. i) The surface of the adsorbent is uniform, that is, all the adsorption sites are equivalent. ii) Adsorbed molecules do not interact. iii) Each site can hold at the most one molecule of adsorbate (mono-layer coverage only). iv) The adsorbed molecules are localized; i.e., they do not move around on the surface. v) Dynamic equilibrium exists between the adsorbed and free adsorbate molecules. The simplest form of mathematical expression for Langmuir adsorption isotherm is $\theta = bp/(1 + bp)$ for adsorption at gas-solid interface and $\theta = bc/(1 + bc)$ for adsorption at solution-solid interface. Theta (θ) represents the fraction of adsorbent surface covered by adsorbate, p & c are the equilibrium pressure and equilibrium concentration, respectively and b the ratio of rate constants for adsorption and desorption

reactions, is the equilibrium constant for the adsorption-desorption equilibrium.
$$\theta = \frac{n}{n_m} \text{ or } \frac{n}{n_m} = \frac{bc}{1+bc} \text{ or } n = \frac{n_m bc}{1+bc} = \frac{ac}{1+bc}$$

In the above equation, n is the number of moles adsorbed per gram of adsorbent, n_m is the maximum number of moles that the surface can take up under identical conditions and a = n_mb is another constant. The monolayer capacity of adsorbent (n_m) is defined as the number of moles of adsorbate required to cover the surface of one gram of adsorbent by a monolayer. The amount adsorbed, n and the monolayer capacity n_m can also be expressed in terms of x/m, the grams of substance adsorbed per gram adsorbent. The Langmuir adsorption isotherm expression can then be written as x/m = a'c/(1 + bc), a' is a constant for a given adsorbate-adsorbent system.

Q. 18. How freundlich and Langmuir adsorption isotherms can be verified?

Ans.: The Freundlich adsorption isotherm equation, x/m = K $p^{1/n}$ or x/m = K $c^{1/n}$, applicable for monolayer adsorption, can be verified by taking log on both sides. A plot of log x/m versus log p or log c should give a straight line passing through origin with intercept log K and slope 1/n. In order to verify whether a set of experimental data obeys Langmuir adsorption isotherm equation x/m = a'c/(1 + bc), it is converted to the equation of a straight line. By inversion and rearrangement of above equation, we get

c/(x/m) = 1/a' + (b/a') c

Thus the adsorption isotherm is verified if a plot of c/(x/m) against c gives a straight line.

Q. 19. If Freundlich and Langmuir adsorption isotherm plots are not straight line, what is your conclusion?

Ans.: Non-linear Freundlich and Langmuir adsorption isotherm plots indicate that the adsorption is not monomolecular, it is multimolecular. Deviation from straight line behavior at higher concentrations may also indicate that the concentrations of adsorbate used are beyond the validity of the isotherm equations. At very high concentrations some other factors such as intermolecular interactions between adsorbate molecules may play a role.

Q. 20. What do you understand by multilayer adsorption? What is the common model for multilayer adsorption?

Ans.: In multilayer adsorption, adsorption isotherm does not exhibit saturation limit. After completion of monolayer, multilayer formation starts. The most common type of adsorption model for multilayer adsorption is Brunauer-Emmett-Teller (BET) adsorption isotherm. This is also called type II adsorption isotherm. The adsorption isotherm equation is written as

$$\frac{p}{v(p_0-p)} = \frac{1}{v_m c} + \frac{c-1}{v_m c} \cdot \frac{p}{p_0}$$

where p and p_0 are the equilibrium pressure and the saturation pressure of the gas/vapour, respectively, c is a constant and v_m is the monolayer capacity. Monolayer capacity v_m is the volume of the gas required to cover the surface of the adsorbent by a monolayer.

Type II adsorption isotherms are usually obtained for physical adsorption of gas/vapour molecules on solid surfaces with a wide distribution of pore sizes. This type of adsorption isotherms are characterized by a transition point or knee point at which formation of mono-molecular layer is complete and multi-molecular layer formation starts. BET adsorption isotherm provides a general method for obtaining surface area of solid from adsorption data.

Q. 21. How specific surface area of an adsorbent can be determined from monolayer capacity?

Ans.: Specific surface area is defined as the total surface area of a material available for adsorption per unit mass of adsorbent. The monolayer capacity (n_m), determined using Langmuir or BET adsorption isotherm can be related to the specific surface area (SSA) of the adsorbent as follows.
SSA = $n_m N_A \sigma^0$ where N_A is the Avogadro's number and σ^0 is the molecular area of the adsorbate (area covered by 1 molecule of adsorbate on the surface). The SI units of specific surface area are m^2/g.

Q. 22. What is the major difference in the adsorption on a solid from gas phase and solution phase?

Ans.: The major difference in the adsorption on a solid from gas phase and solution phase is that in gas phase adsorption, only gas is adsorbed while in solution phase adsorption both solute and solvent may get adsorbed on the solid. Moreover, gas phase adsorption is usually multi-molecular while solution phase adsorption is monomolecular.

Q. 23. Distinguish between physical adsorption and chemisorption.

Ans.: In physical adsorption, the forces involved are weak Van der Wall's forces and heat of adsorption is low while in chemisorption the forces involved are same as that in a chemical bond and heat of adsorption is high. Physical adsorption is reversible and no activation energy is involved while chemisorption is irreversible and activation energy is required. Moreover, physical adsorption can be multimolecular while chemisorption is usually monomolecular.

Q. 24. List five important applications of adsorption?

Ans.: 1) For production of high vacuum, a partially evacuated vessel is connected to a container of good adsorbent such as activated charcoal cooled with liquid air. 2) Adsorption phenomenon is utilized in gas masks to adsorb poisonous gases and thus purify air for breathing. 3) Removal of coloring matter from a substance in processes such as de-colorization of cane sugar. 4) In heterogeneous catalysis, the reactant molecules are adsorbed on the surface of solid catalyst. Many solid adsorbents act as catalysts for gas phase reactions, 5) Adsorption is very widely used as a waste water purification technique. Activated charcoal, zeolites, natural clay minerals, silica and alumina gels are some of the commonly used adsorbents for this purpose.

Q. 25. Explain Langmuir-Hinshelwood mechanism of surface reactions.

Ans.: In many gaseous reactions, the activation energy is much smaller for surface reaction as compared to the same reaction in the gas phase. Bimolecular surface reactions are frequently explained on the basis of Langmuir-Hinshelwood mechanism. This mechanism assumes that the two reactant molecules are adsorbed on neighboring sites on the surface of the solid catalyst. The adsorbed reactants then undergo a bimolecular reaction on the surface. The products formed are desorbed from the surface after the reaction since they are only weakly adsorbed.

17. Colloids

Q. 1. How do you distinguish between a true solution, colloidal sol and suspension?

Ans.: A true solution is a homogeneous mixture of two or more substances with particle size of dissolved material less than 1 nm. For example, solution of sugar in water is a true solution. Particles of a true solution cannot be filtered through filter paper and are not visible to the naked eye. A suspension is a heterogeneous mixture in which the solute size is more than 1000 nm. In suspensions the particles are large enough so that they settle to the bottom of the container and can be filtered using filter paper. For example, muddy water. A colloidal solution, commonly called colloidal sol is also a heterogeneous mixture but with solute particle size intermediate between that of a true solution and a suspension, typically between 1 – 1000 nm. A common example of a colloid is milk. Smoke from a fire is also an example of colloidal system in which tiny particles of solid float in air.

Q. 2. Give some examples of the use of colloids in our daily life?

Ans.: Colloids find numerous applications in our daily life. Some of these are mentioned below.
i) They are used as thickening agents in many industrial and consumer products such as tooth pastes, lotions, lubricants etc. ii) A large number of food stuffs such as milk, butter etc. are colloidal in nature. iii) Many medicines are absorbed easily in the system, if they are in colloidal state. Common examples are skin ointments, some injectable drugs with poor aqueous solubility. iv) In the purification of water, colloidal impurities cannot be removed by filtration. Since colloids are charged particles, colloidal impurities can be removed by coagulating them with particles of charge opposite to that of the colloidal particles. v) In the same way smoke

precipitation is employed to control pollution levels. Sewage water and industrial waste water treatment also uses the same principle. vi) Paints and inks used for protective and decorative purposes are generally colloidal in nature.

Q. 3. What are different ways by which colloidal sols can be classified?

Ans.: Colloidal sols can be classified in the following three ways.

i) Classification based on the state of the dispersed phase and dispersion medium: The most common types of colloids in this classification are sols, gels and emulsions. In a colloidal sol, solid is dispersed in liquid while in a colloidal gel liquid is dispersed in a solid. An emulsion is a dispersion of one liquid in another liquid. Typical examples of sol, gel and emulsion in daily life are paint, jelly and hair cream, respectively.

ii) Classification based on the nature of interaction between dispersed phase and dispersion medium: According to this classification, colloids are classified as lyophilic and lyophobic colloids (lyo means solvent, philic means loving and phobic means hating). They are called hydrophilic and hydrophobic colloids when the solvent is water (hydro means water). In lyophilic colloids, there is high affinity between the dispersed phase and the dispersion medium and the sols are quite stable. The dispersed phase and dispersion medium cannot be easily separated; if separated by some means, the sol can be easily reconstituted. Examples include gum, gelatin, starch, proteins etc. In lyophobic sols there is very little affinity between the dispersed phase and dispersion medium. As such the colloids are not very stable and the dispersed phase is easily precipitated out. Once precipitated, the sol cannot be easily reconstituted. Examples include sols of metals and their insoluble compounds like sulphides (e.g. As_2S_3) and oxides (e.g. $Fe(OH)_3$).

iii) Classification based on the type of particles of dispersed phase. This classification includes three major types of colloids: multimolecular colloids, macromolecular colloids and associated colloids. i) In multimolecular colloids a large number of atoms or small molecules (having diameters of less than 1nm) form aggregates with size in the colloidal range. For example, gold sol, sulphur sol. The molecules in the aggregates are held together by Van der Waal forces. ii)

Macromolecular colloids are obtained on dissolving macromolecules of sizes in the colloidal range in a suitable solvent. Examples include naturally occurring macromolecules such as starch, proteins and cellulose. Many synthetic polymers are also included in this category. iii) Associated colloids are substances which at low concentrations behave as strong electrolytes but at higher concentrations form aggregated particles called micelles which have colloidal characteristics. The formation of micelles takes place only above a particular temperature called Kraft Temperature (T_k) and above a particular concentration called the Critical micelle concentration (CMC). On dilution, these colloids revert back to individual ions. Surface active molecules such as soaps and synthetic detergents are common examples of associated colloids.

Q. 4. How emulsions are classified? What is the role of an emulsifier?

Ans.: Emulsions are broadly classified into two types: oil-in-water (O/W) and water-in-oil (W/O). Oil-in-water emulsion is the dispersion of water-immiscible liquid called oil in water while water-in-oil emulsion is the dispersion of water in a water-immiscible liquid. Semisolid emulsions are called creams. For example, vanishing creams are O/W emulsions and cold creams are W/O emulsions. Since emulsions are unstable systems, an emulsifying agent helps in the formation and stabilization of the emulsion. Usually predominantly hydrophilic emulsifiers are suitable for O/W emulsions and predominantly lipophilic (hydrophobic) emulsifiers are suitable for W/O emulsions.

Q. 5. How colloidal sols are prepared?

Ans.: Lyophilic sols are quite stable and can be easily prepared by shaking the lyophilic substance with the dispersion medium .Examples: Gum, Starch, Gelatin, Albumin etc. Lyophobic sols require special techniques for preparation. The methods can be broadly classified as dispersion methods and condensation methods.

Dispersion methods involve breaking down coarse aggregates into particles of colloidal size by mechanical dispersion (e.g., paints, varnishes, dyes etc.), electrical disintegration (e.g., platinum, silver, gold, copper metal sols) or peptization (e.g., freshly prepared precipitates of ferric hydroxide and silver chloride can be converted into sols by shaking with the dispersion medium in the presence of a small amount of electrolyte).

Condensation methods include various chemical reactions such as double decomposition (e.g., As_2S_3 sol), oxidation (e.g., sulphur sol), reduction (e.g., metal sols), hydrolysis (e.g., $Fe(OH)_3$ sol). These reactions when carried out under specific conditions such as temperature, stirring conditions, order of mixing etc. produce colloidal sols.

Q. 6. What precautions should be observed while preparing lyophobic sols?

Ans.: Since lyophobic sols are less stable and are easily precipitated by electrolytes, all glassware used should be thoroughly cleaned and should be free from ionic contaminants which tend to coagulate lyophobic sols.

Q. 7. How arsenious sulphide sol is prepared in the lab?

Ans.: Arsenious sulphide sol is a lyophobic colloid prepared by hydrolysis of arsenious oxide with boiling water followed by passing H_2S gas through the solution. Arsenious oxide (1.25 g), dissolved in 200 mL of distilled water is boiled for about 10 to 15 minutes, stirring all the time. The solution is cooled and filtered under suction pump to remove any undissolved arsenious oxide. It is then diluted to 250 mL with distilled water and a slow stream of pure hydrogen sulphide gas is passed for about 20 minutes. The solution is again filtered and the arsenious sulphide sol formed is collected as filtrate. The excess of hydrogen sulphide is removed by passing a stream of well washed hydrogen or nitrogen gas through the solution for about 20 minutes. Excess H_2S can also be removed by brisk boiling. The prepared sol is quite stable.

Q. 8. What precautions should be observed in the preparation of arsenious sulphide sol?

Ans.: It is necessary to ensure the complete conversion of arsenious oxide to arsenious sulphide and complete removal of H_2S from solution. This is done as follows.

i) In order to ensure the complete conversion of arsenious oxide to arsenious sulphide, a small portion (3 – 5 mL) of sol is added to $BaCl_2$ solution which causes quick coagulation of the sol. The coagulated sol is filtered and more hydrogen sulphide is passed through the filtrate. If no yellow colour or precipitate develops, complete conversion of arsenious oxide can be assumed.

ii) Complete removal of H_2S from the sol is not possible due to partial hydrolysis of arsenious sulphide. Hence when the odour of H_2S from the sol does not decrease any further, the sol is taken to be sufficiently free from H_2S.

Q. 9. How ferric hydroxide sol is prepared in the lab?

Ans.: Ferric hydroxide sol is prepared by hydrolysis of ferric chloride solution. Twenty five milliliter of 2% ferric chloride solution in water is added to 250 mL of boiling water. Boiling is continued and the water lost by evaporation is replaced until a clear dark brown coloured sol is obtained. The hydrochloric acid produced by hydrolysis of ferric chloride is removed by dialysis of prepared sol.

Q. 10. How colloidal sols are purified?

Ans.: The colloidal sols prepared by any of the methods, usually contain electrolytes and other soluble impurities which can destabilize the sols. The sols can be purified by dialysis, ultrafiltration and ultracentrifugation.

Q. 11. How can colloids be purified by dialysis?

Ans.: The principle involved is based upon the fact that the colloidal particles cannot pass through a dialysis (parchment or cellophane) membrane while the smaller size impurities can pass through it. During dialysis the impurities slowly diffuse out of the dialysis bag leaving behind pure colloidal solution.

Q. 12. What factors are responsible for the stability of colloidal sols?

Ans.: All colloidal particles in a given sample are similarly charged and hence they repel each other. This repulsion between similarly charged particles prevents them from aggregating and settling down and is mainly responsible for the stability of colloidal sols.

Q. 13. Why lyophilic colloids are more stable than lyophobic colloids?

Ans.: The presence of charge on colloidal particles is responsible for the stability of both lyophilic and lyophobic sols. However, lyophilic colloids are more stable than lyophobic colloids because of the additional stability provided by the solvation of colloidal particles due to favourable interactions between the dispersed phase and dispersion medium.

Q. 14. Why colloidal particles are charged? What is the origin of charge on colloidal particles?

Ans.: The major factor responsible for the origin of charge on colloidal particles is the preferential adsorption of either positive or negative ions on their surface. Generally the ions adsorbed are common to the colloidal particles. For example ferric hydroxide sol, prepared by the hydrolysis of ferric chloride, is positively charged due to preferential adsorption of Fe^{3+} ions on colloidal particles. Similarly arsenius sulphide sol is negatively charged due to preferential adsorption of sulphide ions. Charge on colloidal particles can also be due to the dissociation or ionization of surface molecules. For example, when sodium salts of higher fatty acids (soaps) dissociate in water, sodium ions pass into the solution while anions have a tendency to aggregate in solution to form negatively charged colloids (micelles).

Q. 15. How a given colloidal sol can be made positively or negatively charged under different conditions?

Ans.: As already explained, colloidal particles generally acquire charge by preferential adsorption of ions common to the colloidal particles from the electrolyte present in solution. For example, in the presence of potassium chloride, silver chloride sol is negatively charged due to the adsorption of chloride ions while in the presence of silver nitrate, it is positively charged due to the adsorption of silver ions. So the same colloid can be made positively or negatively charged by suitable selection of the composition of the mixture.

Q. 16. How can you find out whether a given sol is positively charged or negatively charged?

Ans.: Neutralization of charge on colloidal particles by addition of oppositely charged ions results in coagulation of colloidal sol. This is a simple and quick way to determine whether a given sol is positively charged or negatively charged. A sol is negatively charged if it is coagulated by positively charged ions and vice versa. Moreover, during electrophoresis a positively charged sol will move towards the negatively charged electrode and a negatively charged sol will move towards the positively charged electrode.

Q. 17. What are protective colloids?

Ans.: Protective colloids are generally lyophilic colloids such as gelatin, natural gum, cellulose derivatives which are added in small amounts to a lyophobic colloid to lessen their sensitivity to the precipitating effect of an electrolyte. The protective ability of a lyophilic colloid is usually defined in terms of gold number. Gold number is defined as the minimum amount of a protective colloid in milligrams which prevents a color change from red to violet of 10mL gold sol by the addition of 1 ml of 10% NaCl solution.

Q. 18. What is flocculation? How the flocculation value of a colloid is defined?

Ans.: The phenomenon of the precipitation or coagulation of a colloidal sol by the addition of the excess of an electrolyte is called flocculation. The number of millimoles of an electrolyte required to bring about the coagulation of one liter of a colloidal sol is called its flocculation value. The folocculation value is inversely proportional to the coagulating power of electrolyte.

Q. 19. Which type of sols can be easily coagulated?

Ans.: Since lyophobic sols are less stable than lyophilic sols, they can be more easily coagulated. Lyophobic sols require much lower concentration of electrolyte (0.1–100 mM) for coagulation whereas lyophilic sols require much higher concentration of the order of 1M of electrolyte.

Q. 20. Why flocculation of a sol is carried out by ions of opposite charge?

Ans.: All colloidal particles in a given sample carry either positive or negative charge. Repulsion between similarly charged particles is responsible for the stability of colloidal sols. The charge carried by colloidal particles is neutralized by oppositely charged ions resulting in flocculation of colloidal sol.

Q. 21. Why divalent and trivalent ions are more effective than monovalent ions for flocculation of a sol?

Ans.: The flocculating power of an ion is based on the Hardy Schulze rule according to which the coagulation capacity of an electrolyte depends upon the valency of the flocculating ion. Flocculating ion is the ion carrying charge opposite to the charge on the colloidal particles. The greater the valency of the flocculating ion, greater will be its coagulating power. Therefore, divalent and trivalent ions are more effective than monovalent ions for flocculation of a sol.

Q. 22. What are various ways other than addition of electrolytes by which a colloidal sol can be flocculated?

Ans.: i) Other than the addition of electrolytes, colloidal sols can also be precipitated by the mutual action of sols. When two sols carrying opposite charges are mixed together in suitable proportions, mutual precipitation occurs. ii) Since traces of electrolytes are essential for stability of sols, removal of all of electrolyte by persistent dialysis also results in the coagulation of sol.

Q. 23. What is Tyndall effect?

Ans.: The Tyndall effect is the scattering of light by colloidal particles when a light beam passes through the colloid. The colloidal particles scatter and reflect light making the light beam visible as a bright streak, called Tyndall cone. Tyndall scattering occurs when the dimensions of the particles that are causing the scattering are larger than the wavelength of the radiation that is scattered. The amount of scattering depends on the frequency of light and the density of particles. Blue light (shorter wavelength) is scattered more strongly than red light (longer wavelength).

Q. 24. What is electrokinetic phenomenon?

Ans.: Since colloidal particles and medium carry opposite charges, when an electric field is applied the particles and medium migrate in opposite directions. Accordingly there are two types of electrokinetic properties, electrophoresis and electro-osmosis. In electrophoresis, the particles can move but not the medium. In electro-osmosis the medium can move but not the particles.

Q. 25. What is zeta potential?

Ans.: Colloidal particles are charged mainly due to the preferential adsorption of positively or negatively charged ions on their surface or ionization of surface molecules. The presence of charge gives rise to a potential at the surface of the particle. The charged particles tend to attract oppositely charged ions from solution resulting in the formation of an electrical double layer in the vicinity of the particle. The double layer consists of two parts. The oppositely charged ions in the vicinity of the charged colloidal particles form a compact layer called Stern layer or fixed layer. The influence of surface charge decreases with distance from the particle and therefore in the second part, called diffused part, the oppositely charged ions are loosely associated with the particle. The fall of potential is sharp in fixed part and gradual in diffused part of double layer. The difference of potential in the fixed and diffused parts of the double layer is called zeta potential or electrokinetic potential. Zeta potential is an important quantity which explains the electrical properties of colloids.

18. Potentiometry

Q. 1. What is the difference between a galvanic cell and an electrolytic cell?

Ans.: In a galvanic cell, energy produced due to the chemical reaction (chemical energy) is converted to electrical energy while in an electrolytic cell, electrical energy produced by an external cell is converted to chemical energy.

Q. 2. Define electromotive force (emf) of a cell.

Ans.: Electromotive force (emf) is the potential difference between two electrodes of a galvanic or electrolytic cell. A potential is developed at the two electrodes due to change in their polarity as a result of chemical reaction in the case of a galvanic cell and the presence of external battery in the case of an electrolytic cell. Positively charged electrode is at a higher potential as compared to the negatively charged electrode. The difference of potential of the two electrodes is the electromotive force or emf of the cell.

Q. 3. What is electrical potential?

Ans.: Electrical potential is defined as the work required to bring a unit positive charge from infinity to a specific point against an electric field. The unit of electrical potential is volt (V). $1V = 1$ J/C $= 1$ Nm/C.

Q. 4. How potential develops at an electrode in a cell?

Ans.: Let us consider the example of a galvanic cell where a metal electrode is in contact with solution of its own ions. The two possible reactions that take place are oxidation and reduction of metal. If the metal has a tendency to get oxidized, metal ions go into the solution leaving the electrons to accumulate in the metal, as a result the metal acquires negative polarity with respect to the solution. On the other hand, if the metal has a tendency to get reduced, the positively charged metal ions in solution are deposited on the metal as neural metal atoms and the metal acquires positive polarity. In either case at equilibrium, a potential difference is established between the metal and solution. Since the metal electrode where reduction takes place acquires positive polarity, it will be at higher potential as compared to the negatively charged electrode where oxidation takes place. So there is a difference of potential between the two electrodes as a result of chemical reaction. In an electrolytic cell, on the other hand, polarity is developed at the two electrodes due to the presence of external battery

Q. 5. In which type of cell chemical energy is converted into electrical energy and vice versa?

Ans.: In electrolytic cells, the two electrodes are connected to an outside source (battery) and the available electrical energy is converted to chemical energy. It is the outside source which makes one electrode positive and the other negative. The development of polarity at the two electrodes is responsible for the chemical reaction in the cell. In a galvanic cell, on the other hand, there is no outside source. When the two electrodes are connected, one of the electrodes has tendency to undergo oxidation reaction while reduction takes place at the other electrode and the flow of current is due to the chemical reaction taking place in the cell. So in this case the chemical energy is converted to electrical energy.

Q. 6. What is a redox reaction?

Ans.: In reduction an atom gains electrons and therefore its oxidation number decreases. This means that the positive character of the species is reduced and the negative character of the species is increased. Oxidation is a process in which an atom loses electrons resulting in increase in its oxidation number. In other words, the positive character of the species is increased. Redox reactions are reactions in which one species is reduced and another is oxidized and therefore, there is transfer of electrons from one species to another.

Q. 7. How do you define a cathode and an anode in a cell?

Ans.: A cathode is always an electrode where reduction takes place and an anode is an electrode where oxidation takes place. However, the polarities of cathode and anode are different in galvanic and electrolytic cells.

Q. 8. What is the polarity of the cathode and anode in a galvanic cell?

Ans.: Let us take the example of a metal/metal ion electrode. In a galvanic cell, the reduction reaction at cathode involves deposition of positively charged metal ions from solution on to the electrode as neural metal atoms and the cathode acquires positive polarity. At anode, the metal has a tendency to get oxidized, metal ions formed go into the solution leaving the electrons to accumulate in the metal and as a result the anode acquires negative polarity with respect to the solution.

Q. 9. What is the polarity of the cathode and anode in an electrolytic cell?

Ans.: In an electrolytic cell, cathode has negative polarity and anode has positive polarity. It must be mentioned that in both type of cells reduction always takes place at cathode and oxidation at anode. However, polarities of cathode and anode are different for the two types of cells. In electrolytic cells, the two electrodes are connected to an outside source (battery) in such a way that cathode acquires negative polarity and the reduction reaction involves deposition of positively charged ions from solution on the electrode. The battery makes other electrode (anode) positive and the negatively charged ions from solution get oxidized and deposited as neutral atoms at the electrode.

Q. 10. What is the standard way of writing a cell? Explain with the help of an example.

Ans.: The cell notation is a shorthand representation of the redox reactions taking place in the cell. A double vertical line (||) is used to separate the anode half reaction from the cathode half reaction. This represents the salt bridge. The anode (where oxidation occurs) is placed on the left side and the cathode (where reduction occurs) is placed on the right side of the double vertical line (||). A single vertical line (|) is used to separate different states of matter and a comma is used to separate like states of matter, on the same side. For example: $Fe^{2+}(aq)$, $Fe^{3+}(aq)$ || $Ag^+(aq)$ | $Ag(s)$.

Q. 11. What is function of a salt bridge in an electrochemical cell?

Ans.: A Salt bridge is used to connect the oxidation and reduction half-cells and maintain electrical neutrality within the internal circuit of an electrochemical cell. Salt bridge consists of an inverted U-tube containing saturated solution of an inert electrolyte such as KCl or NH_4NO_3 thickened with starch or agar gel. Cotton balls are inserted on the two ends of the tube to avoid spillage.

Q. 12. What is liquid junction potential?

Ans.: A potential develops at any interface where there is separation of charge. A potential also develops when electrolyte solutions of different composition are separated by a boundary or a salt bridge. The two solutions may contain different ions or same ions at different concentrations. In either case, the cations and anions move in opposite directions and since they have different mobilities they move at different rates making one solution positively charged relative to the other. For example in sodium hydroxide, hydroxide ion moves approximately five times faster than sodium ion. This will result in an electrical double layer of positive and negative charges at the junction of the two solutions. Thus at the point of junction, a potential will develop because of the ionic transfer. This potential is called liquid junction potential or diffusion potential which is non-equilibrium potential. The magnitude of the potential depends on the relative speeds of the moving ions.

Q. 13. How liquid junction potential can be eliminated or minimized?

Ans.: The liquid junction potential can introduce an error in the measurement of the emf of an electrochemical cell and therefore its effect should be minimized as much as possible. The most common method of eliminating liquid junction potential is to place a salt bridge between the two solutions constituting the junction. Excess of ions present in the salt bridge carry almost the whole of the current across the boundary. Therefore, salts for which the rates of diffusion of cation and anion of the salt are about equal are selected for use in salt bridge. Potassium chloride and ammonium nitrate with about equal transport numbers for the respective anion and cation of the salt are the most frequently used salts in salt bridge.

Q. 14. What is the magnitude and sign of liquid junction potential?

Ans.: The magnitude and sign of liquid junction potential depends on the relative mobilities of the moving ions. In other words it depends on the transport or transference numbers of the anion and cation of the electrolyte. If transference numbers of the anion and cation of the electrolyte are nearly same, liquid junction potential is negligibly small. The salts used in the salt bridge such as potassium chloride and ammonium nitrate have nearly same transference numbers for the anion and cation. If the transference number of cation is greater than that of the anion, the liquid junction potential will be negative and in the reverse case it will be positive. Positive values add to the emf of the cell while negative values decrease the emf of the cell.

Q. 15. What is the principle of a potentiometer?

Ans.: The working principle of potentiometer is based on the phenomenon which is stated in terms of the poggendorf's compensation principle according to which the unknown emf is opposed to another known emf till the two are balanced and no current flows through the circuit. The known emf which can be varied is that of a standard cell included in the circuit. The unknown emf is the emf which is to be measured. The known emf is varied till it is equal to the unknown emf. At this stage the galvanometer shows no deflection.

Q. 16. What is standard electrode potential and standard emf of a cell?

Ans.: Standard electrode potential is defined as the potential of a cell measured under standard conditions, that is with all species in their standard states (1 M for solutions, 1 atm for gases, pure solids or pure liquids) and at a fixed temperature, usually 25°C. Potential of a single electrode as well as that of the complete cell are defined in the same way. Standard emf is the difference in the standard electrode potentials and is written as E^0_{cell}. Potential which is not measured under standard state conditions is written as E_{cell}. The two are closely related by Nernst equation described subsequently.

Q. 17. How can we measure the potential of a single electrode?

Ans.: It is possible to measure only the potential difference between two electrodes using a potentiometer. Single electrode potential can be measured by setting up a cell in which the potential of one of the electrodes is known. The electrode with known potential is called reference electrode while the other electrode whose potential is to be measured is called the principle electrode.

Q. 18. What is a reference electrode and what is its role in the cell?

Ans.: A reference electrode is an electrode which has a stable and known electrode potential. Well known examples of reference electrodes include hydrogen gas electrode (Pt / H_2 (g, 1 atm) / H^+ (aq)), calomel electrode (Hg (l)/Hg_2Cl_2 (s)/KCl (aq)), silver-silver chloride electrode (Ag/AgCl (s)/KCl (aq)). The major role of a reference electrode is to complete the cell. Since the potential of a single electrode cannot be measured, it is connected with another electrode of accurately known potential called reference electrode. From the measured emf, the potential of the principle electrode can be calculated.

Q. 19. What is the difference between oxidation potential and reduction potential?

Ans.: The potential for the oxidation reaction is called oxidation potential while that for the reduction reaction is called reduction potential. Oxidation and reduction potentials are also called anodic and cathodic potentials, respectively since oxidation always takes place at anode and reduction takes place at cathode. If forward reaction is oxidation reaction, the backward reaction is always the reduction reaction. For a given reaction, the oxidation and reduction potentials are equal in magnitude but opposite in sign.

Q. 20. What are different types of electrodes? Give examples.

Ans.: Although a large number of electrodes are possible but the more important of these electrodes are grouped into the following types: (*i*) Metal-metal ion electrodes: A metal in contact with an aqueous solution of a soluble salt of the metal. Example: Zn (s)/$ZnSO_4$ (aq). (*ii*) Metal-metal insoluble salt electrodes: A metal in contact with a sparingly soluble salt of the metal which, in turn, is in contact with an aqueous solution of a soluble salt having a common anion with the sparingly soluble salt. Example: Ag (s)/AgCl(s)/KCl (aq). (*iii*) Gas – gas ion electrodes: A gas is bubbled through an aqueous solution containing gas ions and usually platinum metal is included for making electrical contact. Example: Pt/H_2(g)/ H^+ (aq). (*iv*) Oxidation - reduction or redox electrodes: Platinum electrode is in contact with ions of the same substance in two different valence states. Example: Pt/Fe^{2+} (aq), Fe^{3+} (aq).

Q. 21. What are reversible and irreversible cells? Define reversible emf of a cell.

Ans.: A cell is said to be reversible if it satisfies the following criteria. i) The cell reaction stops when an opposing external emf exactly equal to that of the cell is applied ii) The chemical reaction of the cell is reversed and the current flows in opposite direction when the opposing external emf applied is slightly higher than that of the cell. Any cell which does not obey these two conditions is said to be irreversible cell. For example, Daniel cell (emf = 1.1V) is a reversible cell where the chemical reaction, Zn (s) + Cu^{2+} (aq) → Zn^{2+} (aq) + Cu (s) stops when external emf of 1.1V is applied and the cell reaction is reversed when an emf greater than 1.1V is applied. On the other hand, cell Zn (s)/H^+ (aq)/Ag (s) is an irreversible cell; the cell reaction cannot be reversed since one of the products of the cell reaction, hydrogen gas escapes from the reaction system. Reversible emf is the electromotive force of a reversible cell. Maximum emf is produced by the cell when it behaves reversibly.

Q. 22. When we say that an electrode is reversible with respect to a particular kind of ion, what does it mean?

Ans.: An electrode is reversible with respect to a particular kind of ion when the potential of the electrode changes with the concentration of that type of ion. For example, hydrogen gas and quinhydrone electrodes are reversible with respect to hydrogen ions because the potential of these electrodes changes with the concentration of hydrogen ions. Similarly Ag (s)/AgCl (s)/KCl (aq) electrode is reversible with respect to chloride ions.

Q. 23. How potential developed at an electrode is related to the concentration of reactants and products for the cell reaction?

Ans.: Nernst equation relates the measured reduction potential of an electrode or emf of complete cell to the standard potential, temperature and activities (often approximated by concentrations) of the chemical species undergoing reduction and oxidation. According to this equation
$E_{cell} = E^0_{cell} - (RT/nF) \ln Q$
E_{cell} = cell potential, E^0_{cell} = cell potential under standard conditions, R = gas constant, T = temperature (K), n = number of moles of electrons

exchanged in the electrochemical reaction, F = Faraday's constant, Q = Reaction quotient. Reaction quotient is similar to the equilibrium constant, the only difference is that in reaction quotient, the concentrations of products and reactants are at any given stage during the progress of the reaction and not at the time of equilibrium. For example, the reaction quotient Q = ([C]^c[D]^d)/([A]^a[b]^b) for the reaction aA + bB ⇌ cC + dD. The Nernst equation allows us to calculate the voltage produced by any electrochemical cell if the standard emf of the cell and the concentrations of reactants and products are known.

Q. 24. With the help of an example show how electrode reactions and over-all cell reaction can be written?

Ans.: Let us consider the cell Cu(s)|Cu^{2+}(aq) ∥ Ag$^+$(aq)|Ag(s). Since oxidation always takes place at the left hand electrode and reduction takes place at the right hand electrode, the electrode reactions can be written as follows. Cu(s) → Cu^{2+}(aq) + 2$\bar{e}$ at the left hand electrode and 2Ag$^+$(aq) + 2$\bar{e}$ → 2Ag(s) at the right hand electrode. The over-all cell reaction is Cu(s) + 2Ag$^+$(aq) → Cu^{2+}(aq) + 2Ag(s).

Q. 25. Write chemical reactions at the quinhydrone electrode.

Ans.: Quinhydrone is an equimolar mixture of quinone and hydroquinone. Quinhydrone electrode is a type of redox electrode where hydroquinone is oxidized to quinone and quinone is reduced to hydroquinone. The electrode is reversible with respect to hydrogen ions. The oxidation reaction is Hydroquinone (QH$_2$) → Quinone (Q) + 2H$^+$ + 2$\bar{e}$ while the opposing reduction reaction involves the reduction of quinone to hydroquinone. The potential at the electrode, E_Q depends on the ratio of the activity or concentration of quinone and hydroquinone, and also the hydrogen ion concentration. According to Nernst equation

$$E_Q = E_Q^0 - \left(\frac{RT}{2F}\right) \ln ([Q][H^+]^2/[QH_2]) = E_Q^0 - \left(\frac{RT}{2F}\right) \ln ([Q]/[QH_2]) - \left(\frac{RT}{F}\right) \ln [H^+]$$

$$= E_Q^0 - \left(\frac{RT}{F}\right) \ln [H^+] \text{ for the oxidation reaction and}$$

$$E_Q = E_Q^0 - \left(\frac{RT}{2F}\right) \ln([QH_2]/[Q][H^+]^2) = E_Q^0 - \left(\frac{RT}{2F}\right) \ln([QH_2]/[Q]) + \left(\frac{RT}{F}\right) \ln[H^+]$$
$$= E_Q^0 + \left(\frac{RT}{F}\right) \ln[H^+] \text{ for the reduction reaction}$$

E_Q^0 being the standard potential of quinhydrone electrode. Since quinhydrone is an equimolar mixture of quinone and hydroquinone, the ratio $[QH_2]/[Q]$ is unity and the second term on the right hand side in the above expressions is zero.

Q. 26. What is electrochemical series?

Ans.: The electrochemical series is built up by arranging various redox equilibria in the order of their standard electrode potentials (reduction potentials). The most negative E° values are placed at the top of the electrochemical series and the most positive at the bottom. The potentials are measured with reference to the standard hydrogen electrode (SHE), which is taken as a standard and arbitrarily assigned a potential of zero. Standard hydrogen electrode consists of an aqueous solution containing hydrogen in its oxidized form (the hydrogen ion, H^+) at a concentration of one mole per litre maintained at 25° C in equilibrium with hydrogen in its reduced form (hydrogen gas, H_2) at a pressure of one atmosphere. The reduction half reaction is expressed by the equation $2H^+(aq) + 2e^- \rightleftharpoons H_2(g)$. The following inferences can be drawn from the electrochemical series.

(i) The negative sign of standard reduction potential indicates that the given electrode when joined with SHE acts as anode and oxidation occurs on this electrode. For example, standard reduction potential of zinc is – 0.76 volt. This means that when $Zn^{2+}(aq)/Zn$ (s) electrode is joined with SHE, it acts as anode (–ve electrode) and oxidation occurs on this electrode. Similarly, the +ve sign of standard reduction potential indicates that the given electrode when joined with SHE acts as cathode and reduction occurs on this electrode. For example, standard reduction potential of copper is + 0.34V which means that when $Cu^{2+}(aq)/Cu$ (s) electrode is joined with SHE, it acts as cathode (+ve electrode) and reduction occurs on this electrode.

(ii) The substances, which are stronger reducing agents than hydrogen are placed above hydrogen in the series and have negative values of standard reduction potentials. All those substances which have positive values of reduction potentials and placed below hydrogen in the series are weaker reducing agents than hydrogen.

(iii) Metals on the top (having high negative value of standard reduction potentials) have the tendency to lose electrons readily. These are active metals. The activity of metals decreases from top to bottom. The non-metals at the bottom (having high positive values of standard reduction potentials) have the tendency to accept electrons readily. These are active non-metals. The activity of non-metals increases from top to bottom.

Q. 27. How polarity is assigned to the two electrodes which are combined to form a galvanic cell?

Ans.: Whenever two electrodes are combined to form a galvanic cell, we need to know which electrode will have negative polarity and which electrode will have positive polarity. This information is necessary in order to make appropriate connections and write cell reactions. In a galvanic cell, the electrode at which oxidation takes place acquires negative polarity while the other electrode where reduction takes place acquires positive polarity. The general rule is that we look at the standard reduction potentials of the two electrodes in the electrochemical series. Reduction takes place at the electrode which has higher (more positive) electrode potential while oxidation takes place at the other electrode. So the electrode which has more positive electrode potential will be the right hand electrode and the other electrode will be the left hand electrode.

Q. 28. Write down cell reaction for the cell, $Cr \mid Cr^{3+}(aq, 1\ M) \parallel Ag^{+}(aq, 1\ M) \mid Ag$. Also calculate the cell emf from known electrode potentials.

Ans.: In the cell, $Cr \mid Cr^{3+}(aq, 1\ M) \parallel Ag^{+}(aq, 1\ M) \mid Ag$, oxidation will take place at the chromium electrode since it is written on the left hand side while reduction will take place at the silver electrode which is written on the right hand side. Thus even without knowing the standard reduction potentials, polarity can be assigned to the two electrodes.

Anode half-cell reaction: $Cr(s) \rightarrow Cr^{3+}(aq) + 3e^{-}$, $E^{0}_{Cr \rightarrow Cr^{3+}} = 0.73\ V$

Cathode half-cell reaction: $3Ag^+(aq) + 3e^- \rightarrow 3Ag(s)$, $E^°_{Ag+ \rightarrow Ag} = 0.80$ V
Overall cell reaction: $Cr(s) + 3Ag^+(aq) \rightarrow Cr^{3+}(aq) + 3Ag(s)$
$E^°_{cell} = E^°_{Cr \rightarrow Cr^{3+}} + E^°_{Ag+ \rightarrow Ag}$
$= 0.73$ V $+ 0.80$ V $= 1.53$ V

Note: i) If the anode reaction is written as oxidation reaction and the cathode reaction as reduction reaction, the emf of the cell is the sum of the two electrode potentials.

ii) If both anode and cathode reactions are written as reduction reactions, the emf of the cell is the difference between the right hand ($E^°_R$) and the left hand ($E^°_L$) electrode potentials, $E^°_{cell} = E^°_R - E^°_L$.

Q. 29. In which type of cells, there is no net chemical reaction?

Ans.: Cells in which there is no net chemical reaction are called concentration cells. A concentration cell is a limited form of a galvanic cell that has two equivalent half-cells differing only in concentrations of the electrolyte used. In such cells emf is produced not due to a chemical reaction but due to the transfer of matter from the half-cell with a higher electrolyte concentration to the one with lower electrolyte concentration. Concentration cells are of two types: concentration cells without transference where the two electrolyte solutions are not in direct contact with each other and the concentration cells with transference where the transference of ions from one solution to other takes place directly.

Q. 30. Give one example each of a concentration cell with transference and concentration cell without transference. Also write expression for emf of these cells.

Ans.: The concentration cell with transference is the simplest one in which the electrolyte solutions in the two half cells are in direct contact. A commonly used concentration cell with transference is shown below.

Ag, AgCl (s) | HCl (m_2) ‖ HCl (m_1) | AgCl (s), Ag

The emf of the above cell is given by the equation

$$E = 2t_{H^+} + \frac{2.302\, RT}{F} \log \frac{a_2}{a_1}$$

where t_{H^+} is the transference number of hydrogen ion and a_2 and a_1 are the activities of hydrochloric acid solutions of concentrations m_2 and m_1, respectively. The electrodes are reversible with respect to chloride ions. Such a cell can be used to determine the transport number of other ion of the electrolyte, the hydrogen ion.

In the concentration cell without transference, the solutions in the two half cells are not in direct contact. A typical example is given below. The two end electrodes containing HCl solutions of different concentrations are connected through Ag, AgCl (s) | HCl electrode.

Pt, H_2 (g) | HCl (m_2) | AgCl (s), Ag, AgCl (s) | HCl (m_1) | H_2 (g), Pt

The emf of the cell can be written as

$$E = \frac{2 \times 2.303\, RT}{F} \log \frac{a_1}{a_2}$$

where a_2 and a_1 are the activities of hydrochloric acid solutions of concentrations m_2 and m_1, respectively.

Q. 31. How emf of an electrochemical cell is related to the standard free energy change and equilibrium constant for the cell reaction?

Ans.: Standard free energy change, $\Delta G^0 = -nF\, E^0_{cell} = -RT \ln K$
Where n = the number of electrons involved in the balanced redox reaction, F = Faraday's constant, E^0_{cell} = standard emf of the cell, R = Gas constant, T = Absolute Temperature and K = equilibrium constant for the cell reaction.

Q. 32. How thermodynamic parameters for the cell reaction can be determined from electromotive force measurements?

Ans.: The three basic thermodynamic parameters in physical chemistry are free energy change (ΔG), enthalpy change (ΔH) and entropy change (ΔS). These thermodynamic parameters can be determined by measuring the standard emf of a cell and its temperature-dependence using the following relations.
Standard free energy change, $\Delta G^0 = -nF\, E^0_{cell}$

Standard enthalpy change, ΔH^0 can be obtained from the relation
$$E^0_{cell} = -\Delta H^0/nF + T\,(\partial E^0_{cell}/\partial T)_P$$
The derivative $\partial E^0_{cell}/\partial T$ is determined as the slope of E^0_{cell} versus T plot. Standard entropy change, ΔS^0 can be obtained from the basic relation $\Delta G^0 = \Delta H^0 - T\,\Delta S^0$

Q. 33. What are major applications of potentiometry?

Ans.: Following are some of the major applications of emf measurements using a potentiometer.
i) Determination of solubility product constant of a sparingly soluble salt.
ii) Determination of pH of an aqueous solution. iii) Determination of valency of the metal ions of the electrolyte. iv) Determination of thermodynamic parameters and equilibrium constant of the cell reaction. v) Potentiometric titrations.

Q. 34. How pH value of an aqueous solution can be determined potentiometrically using hydrogen gas electrode?

Ans.: To determine the pH of a solution, the primary requirement is that the principle electrode should be reversible with respect to hydrogen ions, that is, its potential should vary with the concentration of hydrogen ions. The most common example is the standard hydrogen gas electrode. The reduction reaction at the hydrogen electrode is H^+ (aq) + $e^- \rightarrow \frac{1}{2} H_2$ (g) (1 atm) and its potential given by Nernst equation is
$$E_{el} = E^0_{el} + (RT/F) \ln [H^+] = (2.303\,RT/F) \log [H^+]$$
$$= -(2.303\,RT/F)\,pH = -0.0591\,pH \text{ at } 25^0\,C.$$
The standard potential of the standard hydrogen gas electrode is zero by convention. Thus the potential of the hydrogen electrode depends on the pH of the solution. The hydrogen gas electrode is combined with a reference electrode such as calomel electrode of known potential (0.2422) and the complete cell is represented as

Pt | H_2 (1 atm) | H^+ (aq) (Unknown conc.) ‖ KCl (satd. Soln.) | Hg_2Cl_2 (s) | Hg (l)

The emf of the cell, determined potentiometrically is given by

$E_{cell} = E_{calomel} - E_{H^+/H_2} = 0.2422 - (-0.0591\ pH)$
and $pH = (E_{cell} - 0.2422)/0.0591$.

Q. 35. How pH value of an aqueous solution can be determined potentiometrically using quinhydrone electrode?

Ans.: Quinhydrone electrode is very frequently used in lab for potentiometric determination of pH since hydrogen gas electrode is difficult to set up. A pinch of quinhydrone is dissolved in the solution and platinum electrode is employed for making electrical contact. The quinone-hydroquinone redox system involves the following equilibrium.
$C_6H_4O_2$ (Quinone) + 2 H^+ + 2 e^- ⇌ $C_6H_6O_2$ (Hydroquinone) which can be abbreviated as
$Q + 2 H^+ + 2 e^- \rightarrow QH_2$
According to Nernst equation, the potential developed at the electrode can be written as

$$E_Q = E_Q^0 - \left(\frac{RT}{2F}\right) \ln ([QH_2]/[Q][H^+]^2) = E_Q^0 - \left(\frac{RT}{2F}\right) \ln ([QH_2]/[Q]) + \left(\frac{RT}{F}\right) \ln [H^+]$$

$$= E_Q^0 - \left(\frac{RT}{2F}\right) \ln ([QH_2]/[Q]) + \left(\frac{2.303RT}{F}\right) \log [H^+]$$

$$= E_Q^0 - \left(\frac{RT}{2F}\right) \ln ([QH_2]/[Q]) - \left(\frac{2.303RT}{F}\right) pH\ for\ the\ reduction\ reaction$$

E_Q^0 is the standard reduction potential of quinhydrone electrode. Since quinhydrone is an equimolar mixture of quinone and hydroquinone, the ratio $[QH_2]/[Q]$ is unity and the second term on the right hand side is zero.

Therefore,
$$E_Q = E_Q^0 + \left(\frac{2.303RT}{F}\right) \log [H^+] = +0.6996 + 0.0591\ \log [H^+]\ at\ 25^0\ C$$

Quinhydrone electrode is combined with saturated calomel electrode to form a cell. The complete cell can be written as

Hg (l) / Hg_2Cl_2 (s) | KCl (satd. Soln.) ‖ H^+ (unknown conc.) | Q, QH_2, Pt

Since the standard potential of quinhydrone electrode is more positive than that of calomel electrode, quinhydrone electrode will be the right hand electrode and reduction will take place at this electrode. The emf of the cell is given by

$$E_{cell} = E_{Quinhydrone} - E_{Calomel} = +0.6996 - 0.0591\, pH - 0.2422 \text{ or}$$

$$pH = \frac{0{,}6996 - 0.2422 - E_{cell}}{0.0591} \text{ at } 25^0 C.$$

Q. 36. Which types of titrations are carried out potentiometrically?

Ans.: Generally the following three types of potentiometric titrations are carried out.
i) Acid – base titrations ii) Oxidation – reduction (Redox) titrations and iii) Precipitation titrations

Q. 37. Describe briefly how acid-base titrations can be carried out potentiometrically?

Ans.: In acid-base titrations, the concentration of hydrogen ions decreases when an acid is neutralized by a base. Therefore, we need to set up a cell in which the potential of the principle electrode varies with the concentration of hydrogen ions. That is the principle electrode should be reversible with respect to hydrogen ions. Since hydrogen gas electrode is not easy to set up, quinhydrone electrode is frequently used as the principle electrode which is combined with calomel reference electrode to complete the cell. The cell and the relation of E_{cell} to the concentration of hydrogen ions has already been given in Q. 35.

Usually a known volume of acid is taken in the beaker and base is gradually added from the burette. The emf of the cell is determined after each addition. The emf readings are then plotted against the volume of titrant added and the resulting curve is called potentiometric titration curve. Since titration is accompanied by decrease in the concentration of hydrogen ions, the potential of quinhydrone electrode and the emf of the cell decreases on successive addition of base in accordance with the relations

$$E_Q = E_Q^0 + \left(\frac{2.303 RT}{F}\right) \log [H^+] = +0.6996 + 0.0591 \log [H^+]$$

$$E_{cell} = E_{Quinhydrone} - E_{Calomel} = +0.6996 + 0.0591 \log [H]^+ - 0.2422 \text{ at } 25^0 C$$

These relations show that the decrease in the emf of the cell with decrease in the concentration of hydrogen ions is logarithmic which means that the emf of the cell would decrease by 0.0591V for every ten-fold decrease in the concentration of H^+ ions. As such the potetiometric titration curve will show inflexion and the point of inflexion (where the curve is steepest) that is where the emf decreases most rapidly corresponds to the end point of the titration. For strong acid-strong base titration, the end point can be easily located since the curve is almost vertical near the end point. When the acid or base is weak, a derivative plot has to be drawn to locate the end point precisely. The derivative plot is a plot of change in emf per unit change in volume, dE_{cell}/dV against the volume of base added. The maximum in the plot gives the end point of the titration.

Q. 38. How polybasic acids are titrated potentiometrically?

Ans.: In case of polybasic acids there is more than one ionizable hydrogen ion. The hydrogen ions are produced by step-wise dissociation of the acid, each step having its own dissociation constant, K_1, K_2 etc. During potentiometric titration each step will give rise to an inflexion point on the titration curve. For a dibasic acid such as oxalic acid, the two inflexion points can be clearly distinguished on the titration curve since the first dissociation constant, K_1 is at least 100 times greater than the second dissociation constant, K_2.

Q. 39. What difficulty is encountered in the titration of a tribasic acid such as phosphoric acid and how it is overcome?

Ans.: The reasoning discussed above can be extended to a tribasic acid such as phosphoric acid. However, the third dissociation step of phosphoric acid is very weak, the dissociation constant K_3 is very small and the corresponding inflexion point is not sharp enough to be determined accurately. To overcome this difficulty, the pH of the solution is lowered by adding solid calcium chloride to the solution after the second end point. Solid calcium chloride reacts with disodium hydrogen phosphate to liberate hydrochloric acid. Since the end point does not depend on the absolute value of pH but the rate of change of change of pH or potential in the case of potentiometric titration, this method works well.

Q. 40. How dilution effect can be minimized during potentiometric acid-base titrations?

Ans.: In any titration, the added titrant results in the dilution of the solution which is being titrated. Let us see how it affects the potentiometric acid-base titrations. The potential of the principle electrode and the emf of the cell are dependent on the hydrogen ion concentration of solution and the concentration of hydrogen ions decreases both by neutralization and dilution. The change in concentration by dilution of solution is undesirable and should be minimized. Therefore, the concentration of titrant is always kept much higher (about 10 times) than that of the acid to be titrated so that the titration can be completed with much smaller volume of titrant thereby minimizing dilution of reaction mixture.

Q. 41. What type of electrode is used in redox titrations? Give an example of potentiometric oxidation-reduction titration.

Ans.: For redox titrations carried out potentiometrically, the principle electrode consists of an inert electrode such as a platinum wire immersed in a solution containing both the oxidized and reduced forms of the same species. Such an electrode acts as an oxidation-reduction electrode. A typical example is the titration of ferrous ammonium sulphate, also called Mohr's salt, against potassium dichromate. A $Pt/Fe^{3+},Fe^{2+}$ electrode is combined with a reference electrode, usually saturated calomel electrode to complete the cell. Acidic potassium dichromate solution is a strong oxidizing agent which oxidizes ferrous sulphate to ferric sulphate. The emf of the cell is a function of the $[Fe^{3+}]/[Fe^{2+}]$ ratio and therefore, when potassium dichromate is added to ferrous ammonium sulphate solution, the emf of the cell increases. At the end point, an inflexion is seen because of the sudden increase in the emf of the cell due to sharp decrease of $[Fe^{2+}]$ concentration.

Q. 42. What are oxidation-reduction indicators and how they are useful?

Ans.: Some organic compounds can exist in both oxidized (O) and reduced forms (R). The conversion of one form to another involves the reaction, O + nH$^+$ + ne$^-$ → R. Applying Nernst equation,

$$E_{O,R} = E^0_{O,R} - \left(\frac{RT}{nF}\right) \ln ([R]/[O][H^+]^n) = E^0_{O,R} + \left(\frac{RT}{nF}\right) \ln ([O]/[R]) + \left(\frac{RT}{F}\right) \ln [H^+]$$

where $E^0_{O,R}$ is the standard potential of the redox system involved. An important property of these compounds is that their oxidized and reduced forms have different colours. Therefore, if the hydrogen ion concentration is kept constant, the ratio [O]/[R] and hence the ratio of the two colours will enable evaluation of the redox potential ($E_{O,R}$).

Another important feature of these substances is that when added in small amounts to any other redox system (for example, Fe^{3+} + e$^-$ → Fe^{2+}), they spontaneously adjust themselves to the same potential as that of the later. Thus potential of any redox system can be obtained by noting the colour shade developed visually or by means of a colorimeter on adding a small amount of a substance of this class. Such substances are called oxidation-reduction indicators since they enable evaluation of redox potential of a system without setting up potentiometric technique. For example neutral red which is red in oxidized form and colourless in reduced form, ferroin which is light blue in oxidized form and red in reduced form.

Q. 43. What are precipitation titrations?

Ans.: Precipitation titrations are titrations which result in the formation of an insoluble precipitate. For example, titration of AgNO$_3$ with NaCl or KCl forms insoluble salt AgCl. The titration is followed by using Ag (s) /Ag$^+$ (aq) electrode reversible with respect to silver ions as the principle electrode. It is combined with saturated calomel reference electrode to complete the cell. In the cell, silver electrode will be cathode with positive polarity while calomel electrode will be anode with negative polarity since the standard potential of silver electrode is higher (0.80) as compared to the calomel electrode (0.2422). The reduction of silver ions will take place at the silver electrode and the emf of the cell will be given by $E_{cell} = E_{silver} - E_{calomel}$. If AgNO$_3$ is taken in the beaker and KCl is added gradually from the burette, the concentration of silver ions and hence the emf of the cell will decrease on successive addition of KCl solution because the added chloride ions combine with silver ions to form insoluble silver chloride which precipitates out and is removed from the reaction mixture. The emf will become almost constant when whole of silver nitrate is consumed. The end point can be located from the point of intersection of two lines in the curve.

Q. 44. How solubility product and solubility of a sparingly soluble salt such as silver chloride can be determined potentiometrically?

Ans.: The simplest type of cell used for this experiment consists of Ag/Ag$^+$ electrode as the right hand electrode where reduction takes place and saturated calomel electrode as reference electrode on the left hand side. When the solution containing chloride ions (taken in a burette) is added to the titration vessel containing a standard solution of silver nitrate, silver ions are removed as insoluble salt AgCl and as a result the emf of the cell will decrease. The decrease will continue till all the silver ions are removed from the reaction mixture. Further addition of chloride ions will not change emf since electrode is sensitive only to chloride ions. When emf of the cell is plotted against the volume of the titrant added, the concentration of silver ions, [Ag$^+$] at the equivalence point can be determined from the point of intersection of the two lines in the titration curve. Since silver chloride produces equivalent amounts of silver and chloride ions and solubility product,
$$K_{sp} = [Ag^+][Cl^-], [Ag^+] = \sqrt{K_{sp}}.$$
Also for a uni-univalent salt such as silver chloride, the solubility product, $K_{sp} = S^2$ where S is the solubility of the salt. In place of Ag/Ag$^+$ and calomel electrodes, Ag, AgCl (s) | HCl electrode can also be used as principle electrode and quinhydrone electrode as reference electrode.

Q. 45. How mean ionic activity coefficient of an electrolyte can be determined potentiometrically?

Ans.: We know that activity (a) and concentration (c) are related to each other by the expression, a = γ c, where γ is the activity coefficient. Since an electrolyte solution contains both cations and anions which cannot be separated, the activity of an electrolyte solution is expressed as mean ionic activity ($a_{\pm}$) and it is related to concentration by mean activity coefficient ($\gamma_{\pm}$). If concentration is expressed in terms of molality m, mean activity, $a_{\pm} = \gamma_{\pm} m_{\pm}$ where $m_{\pm}$ is the mean ionic molality of solute. In general, for an ionic compound $M_x A_y$ that dissociates 100%, $M_x A_y \rightleftharpoons xM^{z+} + yM^{z-}$, z^+ and z^- represent charges on the cations and anions, respectively.

Activity $a = (a_+)^x (a_-)^y = (a_{\pm})^{x+y}$.

Mean ionic activity $a_{\pm} = (a_+^x\, a_-^y)^{1/x+y}$.

Mean ionic molality and mean ionic activity coefficient are defined in the same way as

$m_{\pm} = (m_+^x\, m_-^y)^{1/x+y}$ and $\gamma_{\pm} = (\gamma_+^x\, \gamma_-^y)^{1/x+y}$.

For a uni-univalent electrolyte like HCl,

$a_{\pm} = (a_+\, a_-)^{½},\ m_{\pm} = (m_+\, m_-)^{½}$ and $\gamma_{\pm} = (\gamma_+\, \gamma_-)^{½}$.

To determine mean ionic activity coefficient of an electrolyte potentiometrically, the following concentration cell without transference may be used.

Pt | QH$_2$, Q, HCl (m$_1$) | AgCl (s), Ag, AgCl (s) | Q, QH$_2$, HCl (m$_2$) | Pt

The quinhydrone electrode is reversible with respect to the cations of the electrolyte, hydrogen ions while silver-silver chloride electrode is reversible with respect to the anions of the electrolyte, chloride ions. One of the concentrations of HCl, say m_2 on the right hand side is kept fixed while m_1 is varied. The measured emf of the cell can be written as

$$E_{cell} = \frac{2 \times 2.303\, RT}{F} \log \frac{(m_{\pm} \gamma_{\pm})_2}{(m_{\pm} \gamma_{\pm})_1}$$

By rearrangement of this equation and making use of Debye-Huckel limiting law,

$\log \gamma_{\pm} = - A |Z_+\, Z_-| I^{½}$

where Z_+ and Z_- are charges on the cation and anion of electrolyte, respectively and I is the ionic strength of solution, the mean ionic activity coefficient of HCl at different concentrations can be calculated.

Q. 46. How can we study the effect of ionic strength on meal ionic activity coefficient?

Ans.: The procedure used is same as explained in the previous question number 45. The only difference is that the hydrochloric acid concentration is kept fixed on both sides of the cell but on one side the ionic strength is varied by the addition of increasing amounts of concentrated solution of potassium nitrate. The data is verified using Debye-Huckel limiting law according to which

$$\log \gamma_\pm = -A |Z_+ Z_-| I^{1/2}$$

where Z_+ and Z_- are charges on the cation and anion of electrolyte, respectively and I is the ionic strength of solution.

$$I = \tfrac{1}{2} \Sigma c_i Z_i^2$$

c_i and Z_i are the concentration and charge on the ion, respectively. The summation is carried out over all the ions present in solution. A straight line plot of $\log \gamma_\pm$ against $I^{1/2}$ with negative slope and no intercept verifies the data.

19. Chemical kinetics

Q. 1. Define rate of a reaction?

Ans.: The rate of a reaction tells us how fast a reaction takes place. Since in a reaction, the reactants are consumed and products are formed, the rate of a reaction can be defined as the decrease in the concentration of reactants per unit time or increase in the concentration of products per unit time. For the reaction $A + B \rightarrow C$, rate $= -d[A]/dt = -d[B]/dt = +d[C]/dt$. The numerical value will be positive in each case since the concentration of reactants A and B decreases with time while that of the product C increases with time. If the stoichiometric coefficients are not unity, for example reaction, $A + 3B \rightarrow 2D$, the concentration of B decreases three times more rapidly than that of A and the concentration of product D increases two times more rapidly than the decrease in concentration of A. In such cases each change in concentration is divided by the appropriate coefficient. Rate $= -d[A]/dt = -\frac{1}{3} d[B]/dt = +\frac{1}{2} d[D]/dt$.

Q. 2. What is instantaneous rate or initial rate?

Ans.: The rate of a reaction is not a constant quantity. The rate of a reaction decreases with time since the reactants are consumed as the reaction progresses. The instantaneous rate or initial rate is the highest rate observed at the beginning of the reaction when the concentration of reactants is maximum. It is the rate at time t close to zero.

Q. 3. What are the factors on which rate of a reaction depends?

Ans.: Rate of a reaction varies widely with the nature of the reaction. In this context reactions may be broadly classified as slow reactions, fast reactions and moderate reactions. In general slow reactions may take a very long time ranging from days to months for completion. The reaction between hydrogen and oxygen to form water or the reaction between carbon and oxygen to form carbon dioxide are some examples of slow reactions. Fast reactions are complete in seconds or fraction of a second. Examples include some ionic reactions such as the neutralization of acids and bases or precipitation of silver chloride on mixing silver nitrate and sodium chloride. Reactions that are neither fast nor slow are termed as moderate reactions. For example the hydrolysis of an ester such as ethyl acetate to form acetic acid and ethyl alcohol is a moderate reaction. The number of reacting species, their physical state, complexity of the reaction and activation energy barrier involved can greatly influence the rate of a reaction. For a given reaction, rate of a reaction depends on the temperature and the presence of a catalyst.

Q. 4. What is the rate constant of a reaction?

Ans.: The rate of a reaction is proportional to the concentration of reactants raised to some power. The proportionality constant is the rate constant of the reaction. Unlike rate of a reaction, rate constant is a constant quantity it does not depend on the progress of the reaction or the concentration of the reactants. For a given reaction it is a fixed quantity at a specified temperature. For example, if rate, r depends on the first power of the concentrations of reactants A and B, $r \propto [A][B]$, $r = k [A][B]$. The proportionality constant k is the rate constant of the reaction. It is also called specific reaction rate because rate constant is equal to rate if the concentration of reactants is unity. The above expression for rate of a reaction is called rate equation or rate law.

Q. 5. How do you distinguish between differential rate law and integral rate law?

Ans.: For the reaction, $A + B \rightarrow$ Products, rate law written as $-d[A]/dt = k [A] [B]$ is called differential rate law. Differential rate law tells us how rate varies with the concentration of reactants. If we integrate the above equation, we get integral rate law. Integral rate law tells us how concentration of reactants varies with time.

Q. 6. How do you define order of a reaction?

Ans.: The order of a reaction is the power by which the concentration terms in the rate equation are raised. The order can be different w.r.t. each component. For a reaction involving two reactants A and B, rate = k $[A]^m [B]^n$, m is the order w.r.t. component A and n is the order w.r.t. component B. It is not necessarily that m and n are equal to the stoichiometric coefficients of A and B in the balanced chemical reaction. The over-all order is the sum of the orders w.r.t. different reactants. Over-all order = m + n in the above example. The order w.r.t. each component as well as the over-all order need not always be a whole number, it can also be a fraction. For example for the reaction, $H_2 + Br_2 \rightarrow 2\ HBr$, rate = k $[H_2] [Br_2]^{1/2}$. The reaction is first order w.r.t. hydrogen but half order w.r.t. bromine and the over-all order is 1½.

Q. 7. What are elementary reactions and complex reactions?

Ans.: Elementary reactions are those reactions which take place in a single step. Complex reactions, on the other hand take place in more than one step.

Q. 8. Define molecularity of a reaction.

Ans.: Elementary reactions are classified according to their molecularity. The number of reacting species, which are involved in simultaneous collision to bring about a chemical reaction, is called the molecularity of the reaction.

Q. 9. How can we define the order and molecularity of a complex reaction?

Ans.: The order of a complex reaction is the order of the slowest step in the reaction mechanism. Molecularity of a complex reaction has no meaning. Molecularity of each step is different.

Q. 10. Explain with the help of examples, how differential rate laws can be written for reactions in which order with respect to a particular constituent is not equal to the stoichiometric coefficient?

Ans.: If at time t = 0, we start with 'a' moles of reactant A, b moles of reactant B and no products and at time t = t, x moles of reactants are converted to products, the rate laws can be written as follows.
Example 1: Reaction $2A \rightarrow$ Products, is first order in A. Differential rate law: $dx/dt = k(a - 2x)$.
Example 2: Reaction $A + 2B \rightarrow$ Products, is first order in A and first order in B. Differential rate law: $dx/dt = k(a - x)(b - 2x)$
Example 3: Reaction $2A + B \rightarrow$ Products, is second order in A and zero order in B. Differential rate law: $dx/dt = k(a - x)^2$.
Example 4: Reaction $3A \rightarrow$ Products, is first order in A. Differential rate law: $dx/dt = k(a - 3x)$.

Q. 11. How kinetics of complex multi-step reactions can be simplified?

Ans.: Kinetics of a reaction is considerably simplified if the initial concentrations of reactants are kept same. For example, reaction $A + B + C \rightarrow$ Products is first order in A, first order in B and first order in C. Overall it is third order. If we start with a moles of A, b moles of B and c moles of C and x moles of product is formed at time t, then the differential rate law will be $dx/dt = k(a - x)(b - x)(c - x)$ which is difficult to integrate. However, if we start with equal number of moles of the three reactants, say a moles of A, a moles of B and a moles of C, the differential rate law will be $dx/dt = k(a - x)^3$ which is much simpler and easy to integrate to get integral rate law.

Q. 12. What do you understand by reaction mechanism?

Ans.: Complex reactions take place in more than one step. Reaction mechanism is the step by step sequence of elementary reactions by which the over-all reaction occurs.

Q. 13. For the simplest reaction in which a single reactant, A reacts to give products, write differential rate laws for zero, first, second and third order reactions.

Ans.: When a single reactant, A reacts to give products, the first, second and third order reactions can be written as A → Products, 2A → Products and 3A → Products, respectively assuming that the stoichiometric coefficient is equal to the order of the reaction in each case. The differential rate laws are given by $-d[A]/dt = k[A]$, $-d[A]/dt = k[A]^2$, $-d[A]/dt = k[A]^3$ for the first, second and third order reactions, respectively. For the zero order reaction, A → Products, the differential rate law is $-d[A]/dt = k$ since the rate does not depend on the concentration of any reactant.

Q. 14. Write integral rate laws for zero, first, second and third order reactions.

Ans.: For writing integral rate laws we again take examples of simple reactions where a single reactant A reacts to give products and stoichiometric coefficient is equal to the order of the reaction in each case. If at time $t = 0$, we start with 'a' moles of reactant A and no products and at time $t = t$, if x moles of reactants are converted to products, $(a - x)$ moles of reactant A will be left. Accordingly in terms of the concentration of product formed per unit time, the differential rate laws can be written as $dx/dt = k_0$, $dx/dt = k_1 (a - x)$, $dx/dt = k_2 (a - x)^2$, $dx/dt = k_3 (a - x)^3$ for the zero, first, second and third order reactions, respectively. Since we are writing now in terms of products formed, there is a positive sign on the left hand side. These expressions can be integrated to get integral rate laws. The resulting expressions for integrated rate laws are

$k_0 = \left(1/t\right)(a - x)$, $k_1 = \left(1/t\right) \ln (a/(a - x))$, $k_2 = \left(1/t\right)[x/a(a - x)]$

and $k_3 = \left(1/2t\right)[x(2a - x)/a^2 (a - x)^2]$

for the zero, first, second and third order reactions, respectively.

Q. 15. How rate constants for the zero, first, second and third order reactions can be determined from experimental data?

Ans.: From the integrated rate laws, rate constants are usually determined by the graphical methods. The simplest way to use graphical method is to convert the relevant integral rate law expressions to the equations of straight line and then the rate constant can be determined from the slopes of the linear plots. In the integrated rate expressions in Q. 14, $a = c_0$ is the initial concentration of reactant, $(a - x) = c$ is the concentration of reactant at time t and $x = a - (a - x) = c_0 - c$ is the concentration of the product formed. For zero order reaction, $c = k_0 t$ and c versus t plots are linear, i.e. the concentration of reactant decreases linearly with time. Similarly for the first order reaction it can be shown that $\ln c = \ln c_0 - k_1 t$. Logarithm of c ($\ln c$) versus t plots are linear which means that the concentration of reactant decreases exponentially with time. For the second order reaction $[(c_0 - c)]/c_0 c] = k_2 t$ and therefore $[(c_0 - c)]/c_0 c]$ versus t plots should be linear with positive slope. For third order reaction also it can be shown that

$1/2 [(c_0^2 - c^2)/c^2 c_0^2] = k_3 t$ and $1/2[(c_0^2 - c^2)/c^2 c_0^2]$ versus t

plots are linear with positive slope. Knowing c_0 and c from experimental data, the rate constant in each case can be easily determined from the slopes of the linear plots.

Q. 16. Why kinetics experiments should always be carried out in a thermostat?

Ans.: The kinetics of most reactions is highly temperature-dependent and therefore, a constant temperature bath, called thermostat, is an essential requisite for any kinetics experiment. Temperature is always specified with the experimentally determined rate constant.

Q. 17. What are pseudo-first order reactions?

Ans.: A reaction is said to be pseudo-first order if the concentration of all reactants except one remains practically constant during the reaction. If a reactant is present in great excess with respect to the other reactants or if it is a catalyst, its concentration remains practically constant during the course of the reaction and therefore does not affect the rate of the reaction. The concentration (of reactant) which remains practically constant during the reaction can be included in the rate constant of the reaction. For example, in hydrolysis reactions water is one of the reactants. Since water is also a solvent it is present in excess and does not affect the rate of the reaction. Thus hydrolysis reactions are usually pseudo-first order.

Q. 18. What is the order of reaction for hydrolysis of methyl acetate and ethyl acetate?

Ans.: Hydrolysis literally means reaction with water. An ester on reaction with water produces acid and alcohol. Since solvent water is also one of the reactants, it is present in excess and therefore, its concentration remains practically constant. Thus ester hydrolysis is a pseudo-first order reaction.

Q. 19. Why ester hydrolysis reaction is carried out in acidic medium?

Ans.: The ester hydrolysis reaction is usually carried out in acidic medium since acid acts as a catalyst for the reaction.

Q. 20. How rate constant for ester hydrolysis can be determined from concentration versus time data?

Ans.: Since ester hydrolysis is a first order reaction, slope of a plot of logarithm (ln) of concentration of ester at different time intervals versus time is equal to the rate constant of the reaction. However, it is not easy to determine the concentration of ester at different times. Therefore, the concentration of acid formed in equivalent amount as product of the reaction can be easily determined by titration of the reaction mixture with a base. The volume of the base required to titrate the acid which is used as catalyst can subtracted in each case. The corrected volume of base is then used to calculate the concentration of acid formed.

Q. 21. What is half-life time of a reaction and how it is useful?

Ans.: As the name suggests half-life time of a reaction is the time period required for half of the reaction to be completed. The term half-life time is generally found useful to describe any type of decay. For example in nuclear chemistry involving radioactive decay, half-life time refers to the amount of time it takes for half of the radioactive isotope to decay. In medical sciences it refers to the biological half-life of drugs and other chemicals in the human body. In addition, order of a reaction can also be determined from the half-life time values determined at two or more different initial concentrations of reactants.

Q. 22. How half-life time is related to the initial concentration of reactant?

Ans.: The half-life time ($t_{1/2}$) of a reaction can be derived simply by substituting x = a/2 in the integrated rate law expressions in Q. 14. It can be shown that
$$t_{1/2} = a/2k_0, \; t_{1/2} = \ln 2/k_1, \; t_{1/2} = 1/k_2 a, \; t_{1/2} = 3/2k_3 a^2$$
for zero, first, second and third order reactions, respectively. Thus half-life time is directly proportional, independent, inversely proportional and inversely proportional to the square of the initial concentration of reactant for zero, first, second and third order reactions, respectively. In general,
$$t_{1/2} \propto 1/a^{n-1}$$
where n is the order of reaction.

Q. 23. How order of a reaction can be determined experimentally?

Ans.: The following four methods are generally used for determination of the order of a reaction.

i) The use of differential rate law expressions: Differential rate law tells us how rate varies with the concentration of reactants. The general differential rate law expression for a reaction involving a single reactant can be written as

$$dx/dt = r = k_n [c]^n$$

where c is the concentration of reactant at any time t and n is the order of the reaction. The logarithmic form of the above expression, $\log r = \log k_n + n \log [c]$ is the equation of a straight line. A plot of log r versus log [c] gives a straight line with slope n, the order of the reaction.

ii) **The use of integrated rate laws:** Integral rate law tells us how concentration of reactants varies with time. This is a hit and trial method. The experimental data (the concentration at different times) is substituted in the integral rate law expressions and rate constant k is calculated. The constancy of the calculated value of k shows that the assumed order is correct. The verification is usually done graphically by checking the linearity of the relevant plot. For example, the reaction is first order if logarithm of concentration versus time plot is linear.

iii) **The half-life time method:** As already discussed above, the half-life time
$t_{½} \propto 1/a^{n-1}$ where n is the order of reaction. If a reaction is carried out at two different initial concentrations, a_1 and a_2, the ratio of the half-life times
$(t_{½})_1 / (t_{½})_2 = (a_2/a_1)^{n-1}$
Taking log on both sides, it can be shown that the order of the reaction
$n = 1 + \{[\log (t_{½})_1/(t_{½})_2]/[\log (a_2/a_1)^{n-1}]\}$
For a first order reaction, $(t_{1/2})_1 = (t_{1/2})_2$ since half-life time is independent of initial concentration of reactant. So
$(t_{½})_1/(t_{½})_2 = (a_2/a_1)^{n-1} = 1$ and since $a_2 \neq a_1, n - 1 = 0$ or $n = 1$.

iv) **Ostwald isolation method:** The basis of this method is that a reactant present in excess does not affect the rate of the reaction since there is negligible change in the concentration of this reactant with respect to others as a result of the chemical reaction. For a reaction involving more than one reactant, the order with respect to each reactant is determined individually and the over-all order is the sum of orders for various reactants. For example, if there are two reactants A and B, first reactant A is taken in excess and the order with respect to reactant B is determined. Then the experiment is repeated taking reactant B in excess and order with respect to reactant A is determined. The over-all order is the sum of the two orders.

Q. 24. What is activation energy and threshold energy of a chemical reaction?

Ans.: The minimum energy that the reacting molecules must possess in order to react is called the threshold energy. The activation energy is the difference between threshold energy and the energy possessed by reacting molecules. In other words, if the reacting molecules do not have enough kinetic energy, additional energy has to be supplied so that the collision is sufficiently energetic for the chemical reaction to take place. This additional energy is called the activation energy of the reaction.

Q. 25. How does a catalyst affect the activation energy of a reaction?

Ans.: A catalyst provides an alternate path of lower activation energy. The usual answer that 'a catalyst lowers the activation energy' is not a completely correct statement. Because a catalyst does not lower the activation energy for the same path, it provides an alternate path. In general, reactions with high activation energy are slow at ordinary temperatures while those with low activation energy are fast. Thus a catalyst speeds up a reaction by providing another path of lower activation energy.

Q. 26. How activation energy of a reaction can be determined experimentally?

Ans.: The activation energy of a reaction can be determined experimentally by using Arrhenius equation. Arrhenius equation is an empirical equation which gives temperature-dependence of the rate constant of a chemical reaction. According to this equation

$$k = A e^{-E_a/RT}$$

where k, A, E_a, R and T are rate constant, frequency factor, activation energy, gas constant and temperature, respectively. If we take log on both sides we get the equation of a straight line.

$$\ln k = \ln A - E_a/RT$$

The rate constant (k) for a given reaction is determined at different temperatures. Activation energy is obtained from the slope of ln k versus 1/T plot.

Q. 27. How can we study the kinetics of the reaction between peroxydisulphate and iodide ions? How order of reaction can be determined?

Ans.: Peroxydisulphate is a powerful oxidizing agent which oxidizes iodide to iodine.
$$S_2O_8^{2-} + 2I^- \rightarrow 2SO_4^{2-} + I_2$$
The liberated iodine can be estimated by reaction with thiosulphate. To determine the order of the reaction, isolation method is used. In one set iodide is taken in excess and peroxydisulphate concentration is varied while in the other set peroxydisulphate is taken in excess and iodide concentration is varied. The overall reaction is second order, first order with respect to each reactant.

Q. 28. How the ionic strength of solution influences the rate of a reaction? Explain primary salt effect.

Ans.: When reactions involving ionic species occur in solution, addition of an inert salt (which does not take part in the reaction) can speed up the reaction, slow it down or has no effect as compared to when no salt is added. The effect increases with increase in the concentration of added salt. This is because the added salt influences the activity coefficients of ionic reactants and activated complex by increasing the ionic strength of the solution. This is called primary salt effect. The effect of salt on rates of ionic reactions can be understood qualitatively and quantitatively using Bronsted-Bjerrum equation. According to this equation, for a bimolecular reaction between reactants A and B in aqueous solution at 298K
$$\ln k = \ln k_0 + 1.018 \, Z_A \, Z_B \, I^{1/2} \quad (1)$$
k and k_0 are rate constants at ionic strength I and zero, respectively. Z_A and Z_B are the charges on the reactants A and B, respectively and I is the ionic strength of solution. $I = \frac{1}{2} \Sigma c_i Z_i^2$, c_i and Z_i are the concentration and charge on the ion, respectively. The summation is carried out over all the ions

present in solution. The slope of ln k versus $I^{1/2}$ plot tells us about the product of the charges on the reacting species.

The following conclusions can be drawn. If the rate constant increases with increase in ionic strength, the slope is positive which means that either both the species involved in the reaction are positively charged or both are negatively charged. On the other hand, if rate constant decreases with increase in ionic strength, one of the species is positively charged and the other is negatively charged. If the rate constant does not change, it remains constant with increase in ionic strength, one or both of the species involved are neutral.

Q. 29. Give an example of a reaction which can be monitored visually.

Ans.: The reaction between potassium permanganate and oxalic acid in acidic medium is an oxidation – reduction reaction.

$$2\,MnO_4^- + 5\,H_2C_2O_4 + 6H^+ \rightarrow 2\,Mn^{2+} + 10\,CO_2 + 8\,H_2O$$

The reaction can be monitored visually since reduction of permanganate by oxalic acid results in the disappearance of pink color of reaction mixture. Time required for the disappearance of pink color is noted in reaction mixtures prepared with different proportions of the reacting species. The data so obtained is analysed in the usual way.

Q. 30. What are clock reactions? Explain iodine clock reaction.

Ans.: Clock reactions represent a unique class of chemical reactions which demonstrate chemical kinetics in action. In these reactions when colorless reagents are mixed, nothing happens initially but after a certain amount of time called induction period, a physical change such as change of color of the reaction mixture occurs. The most common example is the iodine clock reaction which exists in several variations each involving iodine species (as iodide ion, free iodine or iodate ion) and redox reagents in the presence of starch. Amongst iodine clock reactions, the typical example is the reaction between iodide ion and hydrogen peroxide in acidic medium.

$$2\,H_3O^+ + 2\,I^- + H_2O_2 \rightarrow 4\,H_2O + I_2$$

Hydrogen peroxide is an oxidizing agent that oxidizes iodide ions to iodine in acidic medium. It is a bimolecular reaction and follows the rate equation

$-d[H_2O_2]/dT = k\,[H_2O_2]\,[I^-]$.

Solution of hydrogen peroxide is mixed with one containing potassium iodide, starch and sodium thiosulphate. Iodine produced in the reaction is reduced back to iodide by thiosulphate.

$$2\,Na_2S_2O_3 + I_2 \leftrightharpoons 2\,NaI + Na_2S_4O_6$$

This continues till all the thiosulphate has been consumed. Any further iodine formed in the reaction is free, it combines with starch to form blue starch-triiodide complex. Due to this reason, after sometime the colorless mixture suddenly turns dark blue. The concentrations of each of the reactants can be varied and the time it takes to produce a fixed amount of iodine can be measured. This can be done by adding a fixed amount of sodium thiosulphate to the reaction mixture, which will react with a fixed amount of iodine. When the sodium thiosulphate has been used up the blue colour will suddenly appear in solution.

20. Cryoscopy

Q. 1. What is cryoscopy?

Ans.: Cryoscopy is a technique used for measuring the depression in freezing point of a solvent caused by the dissolution of a solute in the solvent.

Q. 2. Why it is called cryoscopy?

Ans.: The name cryoscopy is derived from greek word, cryo which means icy cold.

Q. 3. Define freezing point. Is it different from melting point?

Ans.: The freezing point is the temperature at which a liquid when cooled starts changing into a solid and melting point is the temperature at which a solid when heated starts changing into liquid. In theory the two terms describe the same temperature and the melting point of the solid should be the same as the freezing point of the corresponding liquid. At this temperature the solid and liquid phases coexist in equilibrium. However, small difference between these quantities can be observed in practice. This is because it is possible to cool some liquids below their freezing points without forming a solid (supercooling) while it is difficult almost impossible to heat a solid above its melting point without converting it to liquid. A liquid can become supercooled because it is difficult for particles to organize themselves in a regular structure that is characteristic of that particular substance. Due to this reason solids generally melt over a very narrow range of temperatures and melting points can be used to identify solids and provide information about their purity.

Q. 4. Why a non-volatile solute lowers the vapor pressure of the pure solvent?

Ans.: The vapour pressure of a liquid is determined by the tendency of the liquid molecules present on the surface to escape to the gas phase. A liquid which evaporates easily will have high vapour pressure. The whole surface of a pure liquid is occupied by solvent molecules which are volatile. The surface of a solution, on the other hand, has fewer solvent molecules available for vaporization since some of the surface is occupied by non-volatile solute molecules. Due to the presence of lesser number of volatile solvent molecules on the surface, the vapour pressure of solvent in solution is lower than that of the pure solvent at the same temperature.

Q. 5. How do you define depression in freezing point?

Ans.: Freezing point is the temperature at which the solid and liquid forms of a substance are in equilibrium. Since the dissolution of a solute in a solvent lowers the vapour pressure of the solvent, a solution freezes at a lower temperature as compared to the pure solvent. The depression in freezing point, ΔT is defined as the difference in freezing point of the pure solvent and solution.

Q. 6. What are colligative properties? Depression in freezing point is a colligative property?

Ans.: Colligative properties are those properties which depend only on the number of solute particles in solution. They do not depend on the nature of the solute species present. However, they do depend on the nature of the solvent. Yes, depression in freezing point is a colligative property.

Q. 7. Name some other colligative properties.

Ans.: Relative lowering of vapour pressure, elevation in boiling point and osmotic pressure are some other colligative properties.

Q. 8. Why a solute depresses the freezing point of the solvent?

Ans.: As already discussed, the vapor pressure of a solution containing a nonvolatile solute is always less than that of the pure solvent at any given temperature. At freezing point the solid and liquid phases coexist in equilibrium and vapour pressure of the liquid is equal to that of the solid. Due to vapour pressure lowering, the vapor pressure of a solution will be equal to that of the solid at a lower temperature as compared to the vapour pressure of the pure solvent. Thus, the freezing point will be lower for a solution than for the pure solvent. This effect is called freezing point depression. The same principle applies to the use of antifreeze, a non-volatile solute which is added to the water in the radiators in cars. The freezing point of water is lowered and the solution remains in liquid state even at subfreezing temperatures.

Q. 9. How depression in freezing point is related to the concentration of solute?

Ans.: Depression in freezing point is directly proportional to the molality of the solute in solution. Depression in freezing point, $\Delta T_f = k_f\, m$. The proportionality constant k_f is called the molal freezing point depression constant or cryoscopic constant and m is the molal concentration or molality of solute. Cryoscopic constant does not depend on the nature of solute and its concentration in solution. It is a constant for a given solvent.

Q. 10. Define molal freezing point depression constant or cryoscopic constant of a solvent.

Ans.: The cryoscopic constant of a solvent is defined as the depression in freezing point of a solution when one mole of solute is dissolved in one kilogram of solvent.

Q. 11. Why molality is preferred over molarity in freezing point depression studies?

Ans.: Molality is the number of moles of solute dissolved in 1kg of solvent (Moles of solute/1 kg of solvent). Molality is preferred over molarity in freezing point depression studies because molality does not depend on temperature. Neither the number of moles of solute nor the mass of solvent are affected by changes in temperature. Molarity on the other hand, changes with temperature since the volume of solution changes as temperature changes.

Q. 12. Why we generally consider non-volatile solutes for freezing point depression studies?

Ans.: Both volatile and non-volatile solutes cause depression in freezing point of the solvent. Non-volatile solutes are generally considered only because of simplicity in the treatment of experimental data. If the solute is also volatile, it will also vaporize along with the solvent and the vapour phase will consist of both solute and solvent making analysis of data more complicated.

Q. 13. What is the value of cryoscopic constant for water and what are the factors on which this value depends?

Ans.: The cryoscopic constant for water is 1.86 K kg mol^{-1}. The numerical value depends only on the nature of the solvent. It does not depend on the nature of solute and its concentration in solution.

Q. 14. How depression in freezing point is related to the molar mass of solute?

Ans.: Depression in freezing point, $\Delta T_f = k_f\, m$. Molality, m = Moles of solute/kg of solvent.
$$\Delta T_f = k_f [w_2/(w_1 M_2)]$$
where w_1, w_2 and M_2 are the weight of solute (kg), weight of solvent (kg) and molar mass of solute in kg mol^{-1}, respectively. Thus depression in

freezing point, determined by dissolving a known weight of solute in a known weight of solvent, enables determination of the molar mass of solute.

Q. 15. The freezing point of a solution that contains 1.00 g of an unknown solute dissolved in 20 g of benzene was found to be 2.48°C. The freezing point of pure benzene is 5.48°C. The cryoscopic constant for benzene is 5.12° kg mol⁻¹. What is the molar mass of the solute?

Ans.: The depression in freezing point, ΔT_f = T_f (Pure benzene) - T_f (Solution) = 5.48 – 2.48 = 3.00.

$\Delta T_f = k_f [w_2/(w_1 M_2)]$ or $M_2 = k_f [w_2/(w_1 \Delta T_f)]$

$M_2 = 5.48 \left[1.00 \times 10^{-3}/(20 \times 10^{-3} \times 3.00)\right] = 0.091 \, kg/mol = 91 \, g/mol$

Note: The units of K_f whether written as ⁰ kg mol⁻¹ or K kg mol⁻¹ mean the same thing since the size of degree is same in centigrade and kelvin scales. ΔT_f will have the same value whether the freezing points of pure solvent and solution are expressed in centigrade or kelvin units.

Q. 16. Define Van't Hoff factor.

Ans.: Solutes which undergo dissociation or association in solution are found to show abnormal colligative properties. Van't Hoff factor, i is a unitless constant directly related to the degree of association or degree of dissociation of the solute in the solvent. Van't Hoff factor, i is defined as

$$i = \frac{Observed\ colligative\ property}{Normal\ colligative\ property} = \frac{Normal\ molar\ mass}{Observed\ molar\ mass}$$

There is inverse relationship between colligative property and molar mass. Van't Hoff factor, i = 1 for substances which do not associate or dissociate (ionize) in solution (e.g. urea, sugar etc). For substances which associate in solution, i < 1. For example, for 100% dimerization of benzoic acid in benzene, i = 0.5. For salts which dissociate (ionize) into ions, Van't Hoff factor, i > 1. For example, Van't Hoff factor, i = 2 for NaCl and i = 3 $MgCl_2$ assuming 100% dissociation of the salts.

Q. 17. How do you define the degree of association and the degree of dissociation of a solute?

Ans.: The degree of association is defined as the fraction of solute which associates and the degree of dissociation is the fraction of solute which dissociates or ionizes in solution.

Q. 18. Can we determine the degree of dissociation of a solute from the depression in freezing point? If so how?

Ans.: Yes, the degree of dissociation of a solute, α can be determined from the Van't Hoff factor obtained from the depression in freezing point measurements. Van't Hoff factor, i > 1 for solutes undergoing dissociation. The degree of dissociation is related to the Van't Hoff factor by the equation,

$$\alpha = \frac{i-1}{n-1}$$

where n is the number of species formed after dissociation.

Q. 19. How can we determine the degree of association of a solute from the depression in freezing point?

Ans.: The degree of association of solute can also be determined from the Van't Hoff factor obtained from the depression in freezing point measurements. Van't Hoff factor, i < 1 for solutes undergoing association. Degree of association,

$$\alpha = \frac{n(1-i)}{n-1}$$

where n is the number of solute molecules that combine to form the associated species.

Q. 20. Why for some solutes, molar mass determined by cryoscopic method is different from the normal molar mass?

Ans.: The reason has already been discussed. This happens if the solute associates, dissociates or reacts chemically with the solvent. In all cases, there is a change in the number of species in solution. Since freezing point depression is a colligative property, change in the number of species is responsible for abnormal freezing point depression measurements and hence molar mass.

Q. 21. Explain Rast method for the determination of molar mass of a solute.

Ans.: The freezing point depression method for determination of molar mass of a solute has been discussed in question 14. In the Rast method, the depression in melting point is determined. We know that the freezing point of liquid phase and melting point of solid phase of any substance are same. Therefore it does not make any difference whether we measure the depression in melting point or depression in the freezing point. This method is used for solutes such as naphthalene, acetanilide which dissolve in molten camphor. The necessary conditions are that the solute should not react with camphor or decompose at its melting point. The basic principle of the method is same as discussed previously for depression in freezing point. It is a micro-method since only a few milligrams of solute is required.

Q. 22. What is the advantage of using camphor in the Rast method?

Ans.: The advantage of using camphor is that the molal depression constant of camphor is very high (about 40 K kg mol^{-1}). This means that dissolution of one mole of solute in one kilogram of camphor results in the depression in freezing point/melting point of camphor by about 40°C. Thus there is no need to use expensive Beckmann thermometer, depression in melting point can be easily read using an ordinary thermometer.

21. Conductometry

Q. 1. Why an electrolyte solution conducts electricity?

Ans.: When an electrolyte is dissolved in water it splits up into charged ions; cations and anions. When an electric field is applied, these charged particles conduct electricity.

Q. 2. What is conductivity and how it is measured?

Ans.: Conductivity is a measure of the ability of an electrolyte solution to conduct electricity. It is defined as the reciprocal of the electrical resistance (R) of a solution placed between two electrodes. Conductivity is measured with a conductivity bridge which is based on the Wheatstone bridge circuit. A Wheatstone bridge is the most common and simplest bridge circuit used for measuring unknown resistance. The bridge has four arms consisting of two known resistances, one variable resistance and one unknown resistance which are connected so as to form a quadrilateral. Voltage is applied between a pair of opposite corners and a galvanometer is connected between the other pair of opposite corners. The bridge is balanced by adjusting the variable resistance so that no current flows through the galvanometer. The ratio of two known resistances is then exactly equal to the ratio of adjusted value of variable resistance and the value of unknown resistance. In this way the value of unknown electrical resistance can easily be measured.

Q. 3. What type of cell is used in a conductivity bridge?

Ans.: The two electrode cell is the most commonly used conductivity cell. It consists of two one centimeter square metal electrodes spaced one centimeter apart enclosed in a glass cover. The electrodes are usually made of platinum because of its high electrical conductivity and resistance to chemical attack. For precise work in highly conducting solutions, platinized platinum electrodes are used. The platinized electrodes are coated with platinum black which is basically a very fine powder of platinum with large surface area. The electrodes are platinized by electro-deposition from an aqueous solution consisting of chloroplatinic acid and lead acetate. Platinization increases the effective area of the electrode surface thereby decreasing current density (current per unit area) and minimizing the polarization effects. For highly conducting solutions, sometimes four electrode cells are also used.

Q. 4. What is cell constant?

Ans.: The cell constant of a cell is a function of the cell dimensions, i.e., the area of the electrodes and the distance between them. It is defined as the ratio of the distance (l) in cm between the electrodes of the conductivity cell and the surface area (A) of the electrodes in cm^2. The units of cell constant are cm^{-1}.

Q. 5. What is specific conductance? How it can be determined from measured conductance?

Ans.: Conductivity or specific conductance is the conductance of 1 cm^3 of solution. Specific conductance (K) is equal to the measured conductance (c) multiplied by the cell constant (l/A).

Q. 6. Why measured conductance is not a definite property?

Ans.: Measured conductance (c) is not a definite property since it varies with the cell dimensions, i.e., the area of the electrode plates and the distance between them. Conductivity or specific conductance, defined as the conductance of 1 cm^3 of solution, is a definite property.

Q. 7. Define molar conductance. How it is related to specific conductance mathematically?

Ans.: Molar conductance (Λ_m) is the conducting power of all the ions produced by one gram mole of a substance. Mathematically, $\Lambda_m = 1000\,K/c$, where K is the specific conductance or conductivity and c is the concentration of the electrolyte in gram moles per liter (molar concentration).

Q. 8. What is the difference between molar conductance and equivalent conductance?

Ans.: Equivalent conductance (Λ_e) is the conducting power of all the ions produced by one gram equivalent of a substance. Mathematically, $\Lambda_e = 1000K/c$, where c is the concentration of the electrolyte in gram equivalents per liter (normal concentration). So molar conductance and equivalent conductance are defined in the same way, the only difference is in the concentration units.

Q. 9. What are the units of conductance, specific conductance, molar conductance and equivalent conductance?

Ans.: Since conductance is defined as the reciprocal of resistance, its units are ohm^{-1}. The cgs units of conductance, specific conductance, molar conductance and equivalent conductance are ohm^{-1}, $ohm^{-1}\,cm^{-1}$, $ohm^{-1}\,cm^2\,mol^{-1}$ and $ohm^{-1}\,cm^2\,gequivalent^{-1}$, respectively. The corresponding SI units of various quantities are conductance: Siemen (S), specific conductance: Sm^{-1}, molar conductance: Sm^2mol^{-1}, equivalent conductance: $Sm^2\,gequivalent^{-1}$.

Q. 10. What are the factors on which conductivity of a solution depends?

Ans.: The conductivity of a solution depends on the following factors.

i) **Temperature:** The conductance of an electrolyte solution increases with increase in temperature due to increase in the mobility of ions and the extent of ionization of the electrolyte.

ii) **Nature of electrolyte and its concentration:** A strong electrolyte furnishes more number of ions in solution as compared to a weak electrolyte due to the difference in their degree of ionization. Therefore, at the same concentration the conductivities of strong electrolytes are much higher than that of weak electrolytes. The effect of the concentration of electrolyte is different for specific conductance and molar conductance. The specific conductance increases with increase in the concentration of solution whereas both the equivalent conductance and molar conductance decrease with increase in concentration. iii) Size and mobility of ions: Larger the size of the ion, the lower is the mobility and hence conductivity. iv) The nature of the solvent and its viscosity: The nature of the solvent affects the extent of ionization of the electrolyte and therefore its conductivity. The ionic mobility and hence conductivity is also reduced in more viscous solvents.

Q. 11. How specific conductance/conductivity varies with the concentration of electrolyte?

Ans.: Specific conductance increases with increase in the concentration of electrolyte. Conductivity or specific conductance is defined as the conductance of 1 cm^3 of solution. Since the number of ions per unit volume of solution increases with increase in concentration of the electrolyte, specific conductance also increases.

Q. 12. How molar/equivalent conductance varies with the concentration of electrolyte?

Ans.: Molar/equivalent conductance of an electrolyte, on the other hand, decreases with increase in the concentration of electrolyte. Molar/equivalent conductance is defined as the conducting power of all the ions produced by one g mole/g equivalent of electrolyte. One g mole/g equivalent produces lesser number of ions in concentrated solution as compared to dilute solution and therefore molar conductance decreases with increase in the concentration of electrolyte. The reason for decreased availability of ions from the same amount of electrolyte is decrease in the degree of dissociation of electrolyte with increase in concentration of electrolyte in the case of weak electrolytes. In the case of strong electrolytes since there are more number of ions per unit volume of solution at higher concentrations, due to attractive electrostatic interactions, oppositely charged ions have a tendency to come together to form ionic doublets and therefore, lesser number of ions are available for carrying current.

Q. 13. Why variation of molar conductance with concentration of electrolyte is different for strong and weak electrolytes?

Ans.: As already discussed in Q.11 and 12, the main reason for the decrease in molar conductance with increase in electrolyte concentration is the decreased availability of free ions both in the case of a weak and strong electrolytes. However, for strong electrolytes there is only weak dependence of molar conductance on the concentration of electrolyte since strong electrolytes are fully dissociated at all concentrations, the decrease in molar conductance with increase in concentration of electrolyte is only due to increased attractive interactions between oppositely charged ions. For weak electrolytes the concentration-dependence of molar conductance is much more pronounced because weak electrolytes are partially dissociated and the degree of dissociation of electrolyte decreases significantly with increase in the concentration of electrolyte.

Q. 14. Why pure water also has some conductance?

Ans.: Water is a weak electrolyte which exists in equilibrium with hydrogen and hydroxyl ions. The presence of hydrogen and hydroxyl ions is responsible for the conductance of water. However, the conductance value is very small because the dissociation constant of water, which determines the concentrations of ions in water, is very low; of the order of 10^{-14} at 25^0C.

Q. 15. What is conductivity water?

Ans.: Since water is termed as a universal solvent, it has a tendency to dissolve diverse kind of impurities. It is therefore, difficult to get 100% pure water. These impurities contribute significantly to the conductance of water. Ideally, conductivity water should be 100% pure water containing no dissolved impurities but practically it is not possible to attain 100% purity. Ordinary laboratory distilled water generally has a conductivity of about 3 -5 x 10^{-4} Sm^{-1}. Most of this conductivity is due to impurities and very little due to ionization of water itself. Conductivity can be reduced to about 4 x 10^{-6} Sm^{-1} by extreme purification using special methods. Water so prepared is called conductivity water.

Q. 16. How water used for conductivity measurements can be purified in the lab?

Ans.: For conductivity measurements triple distilled water should be used. Tap water is first distilled with alkaline potassium permanganate to remove acidic impurities. It is again distilled with concentrated sulphuric acid to remove alkaline impurities. The water so obtained is further distilled in an all glass apparatus. For experiments where precise conductivity value is required in the calculations, conductivity water or specially prepared purified water should be used. In case it is not possible to procure purified water, water of highest purity available can be used and the conductance of water used for preparation of solution should be subtracted from the measured value. However, for experiments such as conductometric titrations it is not necessary to use specially purified water since only relative conductivity values are required.

Q. 17. How ionic conductance is related to ionic mobility?

Ans.: Ionic mobility, μ is defined as the speed of an ion under unit potential gradient. Ionic conductance, λ is related ionic mobility by the expression, $\lambda = Z u F$, where Z is the charge on the ion and F is the Faraday constant. For cations, $\lambda_+ = Z_+ u_+ F$ and for anions, $\lambda_- = Z_- u_- F$. The Faraday constant represents the amount of electric charge (e) carried by one mole or Avogadro's number (N_A) of electrons. It is expressed in coulombs per mole (C/mole). $F = N_A e = 9.65 \times 10^4$ C/mol.

Q. 18. What do you understand by limiting molar conductance?

Ans.: Limiting molar conductance is the molar conductance of the electrolyte at infinite dilution. Molar conductance of both strong and weak electrolytes increases on dilution and tends to acquire a maximum value at infinite dilution which is called limiting molar conductance, Λ^0_m.

Q. 19. How can limiting molar conductance be determined for strong and weak electrolytes?

Ans.: For strong electrolytes, limiting molar conductance Λ^0_m can be obtained from Kohlraush's law

$$\Lambda_m = \Lambda^0_m - Kc^{1/2}$$

Λ_m and Λ^0_m are the molar conductance at any given concentration and at infinite dilution, respectively, K is a constant which typically depends on the stoichiometry of the electrolyte and c in the electrolyte concentration. This is the equation of a straight line and therefore, limiting molar conductance, Λ^0_m can be determined from the intercept of the linear Λ_m versus $c^{1/2}$ plot.

For weak electrolytes, Λ_m versus $c^{1/2}$ plot is not linear and therefore, another form of Kohlraush's law is used. It states that at infinite dilution, the dissociation of the electrolyte is complete, each ion behaves independently and the total molar conductivity is equal to the sum of the conductivities of the cations and anions of the electrolyte. Mathematically,

$$\Lambda^0_m = v_+ \lambda^0_+ + v_- \lambda^0_-$$

v_+ and v_- are the number of cations and anions per formula unit of electrolyte, respectively and λ^0_+ and λ^0_-, the molar ionic conductances of cation and anion, respectively at infinite dilution. Molar ionic conductances of a large number of cations and anions at different temperatures are available in the literature.

Q. 20. Write Debye-Huckel-Onsager conductance equation. For which type of electrolytes it is valid?

Ans.: Debye-Huckel theory is applicable for strong electrolytes. The mathematical expression derived by Debye and Huckel was later modified by Onsagar. The resulting equation is called Debye-Huckel-Onsager conductance equation. According to this equation,

$$\Lambda_m = \Lambda_m^0 - (A + B\Lambda_m^0) c^{1/2}$$

Λ_m is the molar conductance at any concentration and $\Lambda^0{}_m$ is the molar conductance at infinite dilution, called limiting molar conductance. Constants A and B depend on the temperature, coefficient of viscosity and dielectric constant of the medium. The values of $\Lambda^0{}_m$ as well as constants A and B at different temperatures are available in the tables (literature). This is a limiting equation which means that it is applicable at low concentrations because of certain assumptions used in the derivation of this equation. It is strictly applicable for strong uni-univalent electrolytes at low concentrations.

Q. 21. How Debye-Huckel-Onsager conductance equation can be verified?

Ans.: Debye-Huckel-Onsager conductance equation can be verified by plotting molar conductance, Λ_m against square root of the concentration of electrolyte, $c^{½}$. A linear plot with slope = $A + B\Lambda^0{}_m$ verifies the equation.

Q. 22. What is the principle of conductometric titrations?

Ans.: Conductometric titration is a type of titration in which the electrolytic conductivity of the reaction mixture is monitored during the progress of the reaction. Conductometric titrations are based on the principle that during a titration, the conversion of reactants to products is accompanied by change in the number and nature of ions available for current conduction. A conducting ion may either be replaced by another ion having different ionic conductivity or is unable to contribute to conductance due to its low degree of dissociation in the product. An ion which is precipitated out of the reaction mixture will also not be able to conduct. As a result the conductivity of the solution varies during the course of titration. The equivalence point is the point at which there is a sudden transition in the trend of conductivity variation. The equivalence point may be located graphically by plotting the measured conductance as a function of the volume of titrant added.

Q. 23. What is the major source of error in conductometric titrations and how it can be taken care of?

Ans.: Dilution of the reaction mixture during titration is the major source of error in conductometric titrations. The addition of increasing amounts of titrant causes dilution of the reaction mixture which results in change in the concentration of conducting ions and hence the measured conductance values. In order to keep dilution factor to a minimum, the solution in the burette should be several times more concentrated (usually 10 times) than the solution being titrated. A much smaller volume of concentrated titrant solution will then be required for which preferably a micro burette should be used. If this condition is not satisfied, the following equation should be used to take into account the change in volume of reaction mixture during titration.

Actual conductivity = $(v + V)/V$ x Observed conductivity

Here V and v are the volumes of the original solution and the titrant added from the burette, respectively. Dilution factor is also responsible for the non-linearity of titration plots which are difficult to extrapolate.

Q. 24. Explain strong acid-strong base conductometric titration curve.

Ans.: The typical example of a strong acid - strong base conductometric titration is the titration of HCl with NaOH. The reaction HCl + NaOH → NaCl + H$_2$O takes place. If we start with HCl in the titration beaker and NaOH in the burette, before NaOH is added, the conductance is high due to the presence of highly mobile hydrogen ions present in HCl solution. When the base is added, the conductance falls due to the replacement of highly conducting hydrogen ions by the slower moving sodium ions. The hydrogen ions from HCl react with hydroxyl ions from the base to form practically undissociated water. This decrease in the conductance continues till the equivalence point is reached. At the equivalence point, the whole of HCl is consumed and the solution contains only NaCl and water. After the equivalence point, no hydrogen ions are available for the neutralization of added hydroxyl ions, so the conductance increases on further addition of NaOH due to the presence of increasing amounts of highly conducting hydroxyl ions. If dilution of solution is kept minimum, a linear variation in conductance with the volume of titrant added will be observed. The conductometric titration curve will consist of two straight lines, sharp decrease in conductance followed by sharp increase in conductance as increasing volume of titrant is added to the reaction mixture. The point of intersection of the two straight lines gives the end point.

Q. 25. Explain weak acid-strong base conductometric titration curve.

Ans.: The typical example of a weak acid - strong base titration is the titration of acetic acid with sodium hydroxide. The reaction $CH_3COOH + NaOH \rightarrow CH_3COONa + H_2O$ takes place. If we start with CH_3COOH in the titration beaker and NaOH in the burette, before NaOH is added, the conductance is low due to the feeble ionization of acetic acid. When the base is added, there is a further dip in conductance in spite of formation of highly dissociated salt, sodium acetate. This is due to the common ion effect. The degree of dissociation of acetic acid, which is a weak electrolyte, is further suppressed due to the presence of a common ion, acetate ion available from sodium acetate formed. But very soon the conductance increases due to the formation of increasing amounts of sodium acetate which is a strong electrolyte. The hydrogen ions from CH_3COOH react with hydroxyl ions from base to form practically undissociated water. At the equivalence point, the solution contains only CH_3COONa and water. Beyond the equivalence point, conductance increases more rapidly with the addition of NaOH due to the presence of increasing amounts of highly conducting hydroxyl ions. After the initial dip, the conductometric titration curve will exhibit two straight lines, both showing increase in conductance with the volume of the titrant added but with different slopes and the point of intersection of the two straight lines gives the end point.

Q. 26. Explain strong acid - weak base conductometric titration curve.

Ans.: The typical example of a strong acid - weak base titration is the titration of hydrochloric acid with ammonium hydroxide. The reaction HCl + NH$_4$OH $\rightarrow$ NH$_4$Cl + H$_2$O takes place. The first part of the conductometric titration curve is same as in the case of strong acid-strong base titration curve. If we start with HCl in the titration beaker and NH$_4$OH in the burette, before NH$_4$OH is added, the conductance is high due to the presence of highly mobile hydrogen ions. When the base is added, the hydrogen ions from HCl react with hydroxyl ions from base to form practically undissociated water and the conductance falls because highly conducting hydrogen ions are being replaced by the slower moving ammonium ions. This decrease in the conductance continues till the equivalence point. At the equivalence point, the solution contains only NH$_4$Cl and water. But after the endpoint has been reached, the graph becomes almost horizontal since added ammonium hydroxide is a weak electrolyte and its dissociation is further suppressed in the presence of ammonium chloride (common ion effect). The conductometric titration curve will show sharp linear decrease in conductance followed by an almost horizontal line and the point of intersection of the two straight lines gives the end point.

Q. 27. Explain weak acid - weak base conductometric titration curve.

Ans.: The typical example of a weak acid - weak base titration is the titration of acetic acid with ammonium hydroxide. The reaction CH$_3$COOH + NH$_4$OH $\rightarrow$ CH$_3$COONH$_4$ + H$_2$O takes place. The first part of the conductometric titration curve is same as in the case of weak acid-strong base titration curve. If we start with CH$_3$COOH in the titration beaker and NH$_4$OH in the burette, before NH$_4$OH is added, the conductance is low due to the feeble ionization of acetic acid. When the base is added, there is a further dip in conductance in spite of formation of highly dissociated salt, ammonium acetate. This is due to the common ion effect. The degree of dissociation of acetic acid which is a weak electrolyte is further suppressed due to the presence of a common ion, acetate ion available from ammonium acetate formed. But very soon the conductance increases due to the formation of increasing amounts of ammonium acetate which is a strong electrolyte. The hydrogen ions from HCl react with hydroxyl ions from the base to form practically undissociated water. This increase in the conductance continues till the equivalence point. At the equivalence point, the solution contains only CH_3COONH_4 and water. After the endpoint has been reached the graph becomes almost horizontal, since added ammonium hydroxide is a weak electrolyte and its dissociation is further suppressed in the presence of ammonium chloride (common ion effect). After the initial dip, the conductometric titration curve will exhibit two straight lines, increase in conductance followed by an almost horizontal line and the point of intersection of the two straight lines gives the end point.

Q. 28. Explain how the mixture of a strong and weak acid can be titrated conductometrically against a strong base or a weak base?

Ans.: One of the advantages of the conductometric titration method is that the mixture of a strong and weak acid can be analyzed by a single titration. The typical example is the titration of a mixture of hydrochloric acid and acetic acid with sodium hydroxide or ammonium hydroxide. The method is based on the fact that in the acid mixture, due to common ion effect, acetic acid remains practically unionized in the presence of hydrochloric acid; hydrogen ion from hydrochloric acid being the common ion. So at the start of the titration, only hydrochloric acid is titrated; the titration of acetic acid starts only when all the hydrochloric acid has been consumed. Therefore, the conductometric titration curve has two break points. The first break point corresponds to the neutralization of strong acid. The titration of weak acid will start only when the strong acid present in the mixture is completely neutralized. The second break point corresponds to the neutralization of weak acid and after that the conductance increases due to the presence of increasing amounts of hydroxyl ions when strong base is used as the titrant. However, when a weak base is used for titration, the conductance remains almost constant after complete neutralization of the acid mixture and the end point is sharper. Therefore, the acid mixture should preferably be titrated with a weak base rather than a strong base.

Q. 29. What do you understand by the terms degree of hydrolysis and hydrolysis constant?

Ans.: An acid and base always react to form salt and water. The opposite reaction, the reaction of the salt formed with water to reform the acid and base is called hydrolysis reaction of the salt. For example, the hydrolysis reaction of sodium acetate can be written as $CH_3COONa + H_2O \rightarrow CH_3COOH + NaOH$. The degree of hydrolysis is the fraction of the salt which gets hydrolysed. Hydrolysis constant is the equilibrium constant for the hydrolysis reaction. Since water is a solvent it is present in excess and therefore, its concentration remains practically constant. The concentration of water is included in the hydrolysis constant. Hydrolysis constant, K_h can be defined as $K_h = [CH_3COOH][NaOH]/[CH_3COONa]$ for the hydrolysis of sodium acetate.

Q. 30. How hydrolysis constant is related to the degree of hydrolysis?

Ans.: Taking the example of the hydrolysis of sodium acetate again, if c is the concentration of salt and α is the degree of hydrolysis, hydrolysis constant, $K_h = c\alpha^2/(1 - \alpha)$

Q. 31. How hydrolysis constant (K_h) is related to the dissociation constant of a weak acid (K_a) or a weak base (K_b)?

Ans.: Hydrolysis constant is equal to the ratio of the ionic product of water (K_w) and the dissociation constant of weak acid (K_a) in the case of the salt of a weak acid and strong base and the ratio of the ionic product of water (K_w) and the dissociation constant of weak base (K_b) in the case of the salt of a weak base and strong acid.
$K_h = K_w/K_a$ and $K_h = K_w/K_b$

Q. 32. With the help of an example explain how degree of hydrolysis and hydrolysis constant of a salt can be determined conductometrically?

Ans.: When the salt of a weak acid or weak base is dissolved in water, hydrolysis occurs. The conductance of the solution of the salt in water is partly due to the ions of the unhydrolysed salt and partly due to the ions of the acid or base formed by hydrolysis. Let us consider the hydrolysis of aniline hydrochloride which is the salt of a strong acid (HCl) and weak base (aniline, $C_6H_5NH_2$). The hydrolysis reaction can be written as

$$C_6H_5NH_3Cl + H_2O \leftrightarrows C_6H_5NH_3OH + HCl$$

Let c be the molar concentration of the salt and α be its degree of dissociation.

The hydrolysis constant, $K_h = K_w/K_b = c\alpha^2/(1-\alpha)$ (1)

Now the molar conductance of the above equilibrium mixture will be the sum of the conductance of the ionized but unhydrolyzed salt and that of the hydrolyzed salt which yields the ionized strong acid (HCl) and practically unionized weak base ($C_6H_5NH_3OH$). On per mole basis, the amount of unhydrolyzed salt is represented by $(1-\alpha)$ and the free acid by α. The experimentally determined molar conductance (Λ) of the salt solution can then be written as

$$\Lambda = (1-\alpha)\Lambda_c + \alpha \Lambda^0_{HCl} \quad (2)$$

where Λ_c is the molar conductance of unhydrolysed salt and Λ^0_{HCl} is the molar conductance of HCl at infinite dilution which can be obtained from tables (literature) as the sum of the molar conductance of hydrogen and chloride ions assuming complete dissociation of HCl. The molar conductance of the unhydrolyzed salt Λ_c can be determined by adding excess of weak base aniline hydroxide to aniline hydrochloride solution. The added weak base which can be regarded as completely unionized in the presence of its salt, suppresses the hydrolysis of salt (aniline hydrochloride) without affecting its ionization. On rearrangement of equation (2), the degree of hydrolysis of the salt at the given dilution is given by the expression

$$\alpha = \frac{\Lambda - \Lambda_c}{\Lambda^0_{HCl} - \Lambda}$$

Since hydrolysis constant $K_h = c\alpha^2/(1-\alpha)$, it can be calculated from the experimentally determined α value.

Q. 33. How solubility and solubility product of a sparingly soluble salt can be determined conductometrically?

Ans.: We know that the molar conductance, $\Lambda_m = 1000 K/c$, where K is the specific conductance or conductivity and c is the concentration of the electrolyte in gram moles per liter (molar concentration) and solubility (S) is defined as the concentration of saturated solution of a salt. If the salt is sparingly soluble, the saturated solution is also very dilute and its molar conductance (Λ_m) may be taken as approximately equal to the molar conductance at infinite dilution (Λ^0_m). Therefore, we can write $\Lambda^0_m = 1000\ K/S$ or solubility, $S = 1000\ K/\Lambda^0_m$. According to Kohlraush's law of independent migration of ions, $\Lambda^0_m = v_+\ \lambda^0_+ + v_-\ \lambda^0_-$ where v_+ and v_- are the number of cations and anions per formula unit of electrolyte, respectively and λ^0_+ and λ^0_-, the molar ionic conductance of cation and anion, respectively at infinite dilution. Since λ^0_+ and λ^0_- values for different ions are available in the literature, experimental measurement of specific conductance of saturated solution of a sparingly soluble salt enables determination of its solubility (S). For accurate results conductivity of water which is used for preparation of solution should be subtracted from the conductivity of saturated salt solution. For a uni-univalent salt such as silver chloride, the solubility product, $K_{sp} = S^2$.

Q. 34. How solubility product data can be used to determine the thermodynamic parameters for the dissolution process?

Ans.: The standard free energy change for the dissolution process
$\Delta G^0 = -\ 2.303\ RT \log K_{sp}.$

The enthalpy of solution, ΔH^0 can be calculated from the solubility product values at two different temperatures using equation
$$\frac{\log K_{sp}(1)}{\log K_{sp}(2)} = \frac{\Delta H^0}{2.302\ R}\left(\frac{1}{T_2} - \frac{1}{T_1}\right)$$
where $K_{sp}(1)$ and $K_{sp}(2)$ are the solubility products at temperatures T_1 and T_2, respectively. The standard entropy change, ΔS^0 can be obtained from the basic relationship
$\Delta G^0 = \Delta H^0 - T\Delta S^0$

Q. 35. How degree of dissociation of a weak electrolyte can be determined conductometrically?

Ans.: Since the decrease in the degree of dissociation is the primary reason for the decrease in molar conductance of a weak electrolyte with increase in concentration of electrolyte, the ratio of molar conductance at any concentration (Λ_m) to that at infinite dilution (Λ^0_m) is equal to the degree of dissociation (α) of the electrolyte, $\alpha = \Lambda_m/\Lambda^0_m$. As already explained in Q. 19, molar conductance at infinite dilution, $\Lambda^0_m (= \nu_+ \lambda^0_+ + \nu_- \lambda^0_-)$ values for different electrolytes are available in the literature. We know that $\Lambda_m = 1000\ K/c$ and therefore, measurement of specific conductance at any given concentration of electrolyte enables calculation of the degree of dissociation of the electrolyte. The effect of concentration of electrolyte on the degree of dissociation can also be easily demonstrated by making measurements at different concentrations of the electrolyte.

Q. 36. What is Ostwald dilution law and how it can be verified using conductivity measurements?

Ans.: Law of chemical equilibrium can be applied to weak acids and weak bases since they are always in a state of ionization equilibrium. For a uni-univalent acid such as HA, the law can be stated mathematically as
$$K = c\alpha^2/(1-\alpha)$$
where K, c and α are the ionization constant, concentration and degree of dissociation of the acid, respectively. The law can be verified using conductance measurements since the ratio of molar conductance at any concentration (Λ_m) to that at infinite dilution (Λ^0_m) is equal to the degree of dissociation (α) of the electrolyte, $\alpha = \Lambda_m/\Lambda^0_m$ as explained in Q.35 as well. Constancy of K for different concentrations of the acid verifies the law.

Q. 37. What is meant by the strength of an acid and how it can be determined conductometricaly?

Ans.: The strength of an acid is its capacity to furnish hydrogen ions in the solution. The dissociation constant of a weak acid or base is a measure of its strength. As already explained in Q.36, conductometric method can be employed for the determination of the degree of dissociation and dissociation constant of a weak acid/base.

Q. 38. How can we compare the relative strengths of acetic acid and monochloroacetic acid conductometrically?

Ans.: Acetic acid and monochloroacetic acid are both weak acids which dissociate partially depending on their concentration in solution. However, monochloroacetic acid is stronger than acetic acid due to the presence of a strong electron withdrawing chlorine atom which pulls the negative charge of the carboxylate anion towards itself by inductive effect resulting in the stabilization of the conjugate base of chloroacetic acid and weakening of the –O-H bond. On the other hand, the negative charge on the oxygen atom in acetic acid is localized and the conjugate base is less stable. The relative strengths of these weak acids can be determined either by comparing the degree of dissociation of these acids at a given concentration or by comparing their dissociation constants. Conductometric determination of the degree of dissociation and dissociation constant of weak acids has already been explained in Qs. 35 and 36.

Q. 39. Define transport number. How transport number is related to the mobility and molar conductance of ions of the electrolyte?

Ans.: Transport number, also called transference number of an ion (t) is defined as the fraction of the total electrical current carried by that ion in solution. Differences in transport number arise due to difference in the electrical mobility of ions. Transport number is related to mobility (u) and molar conductance (λ) of ions of the electrolyte by the following relationships.

$$t_+ = \frac{u_+}{u_+ + u_-} \text{ and } t_- = \frac{u_-}{u_+ + u_-}$$

$$t_+ = \frac{\lambda_+}{\lambda_+ + \lambda_-} \text{ and } t_- = \frac{\lambda_-}{\lambda_+ + \lambda_-}$$

The subscripts + and - refer to the cation and anion, respectively.

Q. 40. What is a saponification reaction? How kinetics of saponification of ethyl acetate by sodium hydroxide can be studied conductometrically?

Ans.: The alkaline hydrolysis of fats is referred to as saponification which means soap making. However, the term is commonly used for the alkaline hydrolysis of any type of ester. The saponification reaction of ethyl acetate for example, can be written as

$$CH_3COOC_2H_5 + NaOH \rightarrow CH_3COONa + C_2H_5OH$$

Initially the conductance of the reaction mixture is due to the sodium and hydroxide ions from sodium hydroxide only since ethyl acetate does not ionize. After the completion of the reaction, ethyl alcohol does not contribute to conductance and the conductance is only due to ions from an equivalent amount of sodium acetate formed. This is known to be a second order reaction. The kinetics of the reaction is considerably simplified if the initial concentrations of ethyl acetate and sodium hydroxide are adjusted to be the same. The differential rate law can then be written as $-dc_A/dt = k\, c_A^2$ which on integration gives

$$\frac{1}{c_A} = kt + \frac{1}{c_0}$$

where k is the rate constant, c_0 is the initial concentration of ethyl acetate and c_A is the concentration of ethyl acetate at time t. The same equation can also be written in the familiar form

$$\frac{1}{a-x} = kt + \frac{1}{a}$$

where a is the initial concentration of ester, x is the amount reacted and $(a-x)$ is the remaining amount at time t. We must keep in mind that the only contribution to conductance is from sodium hydroxide and sodium acetate; ester and alcohol do not contribute to conductance. If c_0 is the initial conductance, c_∞ is the final conductance and c_t is the conductance at time t then

$a \propto c_0 - c_\infty$, $x \propto c_0 - c_t$ and $(a-x) \propto c_t - c_\infty$.

The rate constant k can be determined by measuring slope of the straight line plot of $1/(c_t - c_\infty)$ against time t.

Q. 41. What are precipitation titrations? How precipitation titrations are carried out conductometrically?

Ans.: Precipitation titrations involve the formation of an insoluble precipitate during titration. Most of these reactions are ionic in nature and can be studied conductometrically. The most common examples are argentometric titrations which involve precipitation of silver halides using silver nitrate as the precipitating agent. Such titrations are generally carried out in the presence of a small amount of alcohol which reduces the solubility of the precipitate formed. Alcohol also helps in flocculation of precipitates which are in suspended state.

The shape of the titration curve can be explained taking example of the titration of KCl with $AgNO_3$. Initially the conductance is due to KCl which is a strong electrolyte, addition of $AgNO_3$ results in precipitation of AgCl and formation of KNO_3. Chloride ions are replaced by nitrate ions and since the ionic conductance value of chloride and nitrate ions is comparable, there will not be any effective change in conductance of solution. However, when all the chloride ions are consumed, added silver nitrate will result in sharp increase in conductance due to the presence of increasing amount of silver and nitrate ions. The point in the conductometric titration curve where the initial horizontal line and the line showing sharp increase in conductance intersect is the equivalence point.

Q. 42. Which factors can cause errors in precipitation titrations?

Ans.: Following factors may cause errors in location of end points in precipitation titrations.

i) Adhering of precipitates to the electrode surfaces. ii) Occlusion or trapping of ions from solution within the precipitates. iii) The precipitation reaction must be sufficiently rapid and complete. Incomplete or slow precipitate formation leads to errors. iv) The precipitate formed must settle down. Suspension of precipitates in solution may cause errors in conductance measurements. This is usually taken care of by adding a small amount of alcohol to the reaction mixture.

22. Polarimetry

Q. 1. What are enantiomers/optical isomers?

Ans.: Different compounds with the same molecular formula but different chemical structures are called isomers. Stereoisomers are the type of isomers that have the same bond connectivity that is they consist of same atoms connected in the same sequence but have different arrangement in three-dimensional space. Stereoisomers that are non-superimposable mirror images of one another (i.e., chiral) are called enantiomers or optical isomers. For example, our left hand is a mirror image of the right hand. Optical isomers also have no axis of symmetry, which means that there is no line that bisects the compound such that the left half is a mirror image of the right half. Enantiomers have identical chemical and physical properties except for the ability to rotate plane-polarized light by equal amounts but in opposite directions.

Q. 2. What is a racemic mixture?

Ans.: An asymmetric carbon atom is one that is bonded to four different atoms or groups. The configuration of such a molecular unit is chiral. A chiral molecule is non-superimposable on its mirror image and the structure may exist in either a right-handed configuration or a left-handed configuration. If only one enantiomer is present the sample is considered as optically pure. A racemic mixture has equal amounts of left- and right-handed enantiomers of a chiral molecule. Since a racemic mixture is 50:50 mixture of enantiomers with equal and opposite optical rotation, it is always optically inactive.

Q. 3. What are meso compounds?

Ans.: Meso compounds are stereoisomers that are superimposable on their mirror images. They are considered achiral and optically inactive. A meso compound has chiral centers but also has an internal plane of symmetry which renders the molecule optically inactive.

Q. 4. What are stereoisomers and diastereomers?

Ans.: There are two types of stereoisomers; enantiomers and diastereomers. Enantiomers contain chiral centers and are non-superimposable mirror images of one another. Diastereomers also contain chiral centers and are non-superimposable but are not mirror images. While enantiomers exist in pairs, many diastereomers are possible for a given molecule depending on the number of stereocenters. For example, the cis/trans isomers of 2-butene are diastereomers. Diastereomers also show optically activity but they can have different physical properties and reactivity.

Q. 5. What is optical activity?

Ans.: Optical activity is the ability of a chiral molecule to rotate the plane of polarized light. The experimental parameter is the angle of rotation, defined as the number of degrees by which the solution of a substance rotates the plane of polarized light. It is measured using a polarimeter. Glucose, fructose, lactic acid, tartaric acid, are some of the examples of optically active compounds.

Q. 6. How optical activity is measured?

Ans.: Optical activity is measured using a polarimeter. A simple polarimeter consists of a light source, polarizer, sample tube and analyzer. The light source is usually a mercury or sodium-vapour lamp. The polarizer and analyzer are two Nicol prisms, the polarizer is fixed while the analyzer can be rotated. The sample tube is placed between the polarizer and analyzer. The unpolarized light produced by the light source is converted to plane polarized light by passing through the polarizer which is an optical filter. The plane polarized light thus obtained consists of light with waves that oscillate in one direction only. The polarizer and analyzer are initially set at an angle of 90° towards each other so that no light reaches the detector. When plane polarized light is passed through optically active substance in the sample tube, the light waves are rotated either in the clockwise or counterclockwise direction depending on the type of enantiomer. The angle through which the analyzer has to be rotated to get the point of complete darkness of the field of view that is zero transmission is the angle of rotation which is the experimentally determined parameter of optical activity.

Q. 7. What is the reference liquid used while setting up a polarimeter?

Ans.: The reference liquid used while setting up a polarimeter is usually water because pure water is easily available and is not optically active. It does not rotate the plane of polarization of light. Before starting the experiment, the sample tube is filled with water and the angle through which the analyzer has to be rotated to get the position of **complete darkness of the field of** view is the zero reading of the instrument.

Q. 8. Why we need plane polarized light for measuring optical activity? Do chiral molecules affect only polarized light and not unpolarized light?

Ans.: Chiral molecules affect both polarized and un-polarized light. In un-polarized light since the rays have no particular orientation with respect to one another the effect cannot be observed or measured. When plane polarized light is used, the orientation of rays before passing through the chiral substance is known and so the rotation after passing through the sample can be observed and measured.

Q. 9. Why some substances show optical activity whereas others do not?

Ans.: Each individual molecule of any given compound (Chiral or non-chiral) has the ability to rotate the plane of polarized light by a certain amount. The extent and direction of this rotation depends on the orientation of the molecule in relation to the beam. However, optical activity is not a property of individual molecules but a bulk property. For symmetrical molecules such as methane, ethylene, chloroethane, these individual rotations cancel out and thus no optical activity is observed. For compounds that exhibit chirality (enantiomers), for example the sample of a pure enantiomer or one that is enantiomerically enriched, this phenomenon of exact cancellation of individual rotations cannot occur and therefore the result is optical activity. The necessary and sufficient condition for a molecule to be optically active is chirality or dissymmetry, that is molecule and its mirror image must be non-superimposable. It may or may not contain an asymmetric carbon atom. This same cancellation effect is responsible for the optical inactivity of a racemic mixture.

Q. 10. Why sugar is optically active while carbon dioxide and water are not optically active?

Ans.: As discussed in detail in the previous question, due to the presence of an asymmetric center in the sugar molecule, its mirror images are non-superimposable and sugar is optically active while water and carbon dioxide are symmetric molecules with superimposable mirror images and thus they are optically inactive.

Q. 11. What do you understand by the terms dextro-rotatory and levo-rotatory?

Ans.: The optical isomers (enantiomers) interact differently with the plane-polarized light, resulting in clockwise or anticlockwise change in the direction of its propagation. An enantiomer that rotates plane-polarized light in the positive or clockwise direction is called dextrorotatory [(+) or d-], while an enantiomer that rotates the light in the negative or anticlockwise direction is called levorotatory [(-) or l-]. For example, sucrose is dextrorotatory while fructose is levorotatory. It is the main property used to quantify the chirality of a molecular species. If a chiral molecule is dextrorotary, its enantiomer will be levorotary and vice-versa. The enantiomers rotate polarized light by the same number of degrees but in opposite directions.

Q. 12. What is the difference between (R)/(S) system and (+)/(-) system for classification of enantiomers?

Ans.: The $(R)/(S)$ system is based on the absolute configuration of individual molecules. (R) and (S) describe the absolute stereochemistry of chiral centres while the (+)/(-) system describes the direction of rotation of plane polarized light and is based on the macroscopic behavior of a large collection of molecules. The most complete description of an enantiomer combines aspects of both systems. For example, the enantiomer 2-bromobutane is best described as (S)-(-)-2-bromobutane. It is the (S) enantiomer because of its structure and the (-) sign indicates that the enantiomer is levorotatory, it rotates the plane-polarized light in the anticlockwise direction. The sign of optical rotation is not correlated to the absolute configuration.

Q. 13. What are the factors on which angle of rotation depends?

Ans.: The measured angle of rotation is dependent on several factors such as concentration of the sample (or density of pure liquid), length of the sample tube and wavelength of the light passing through the sample. It also depends on the temperature and the solvent used but to a lesser extent. Rotation is given in +/- degrees, depending on whether the sample has d- (positive) or l- (negative) enantiomer.

Q. 14. Define specific rotation.

Ans.: Specific rotation corrects the optical rotation for the cell dimensions and the concentration. It is defined as the angle of rotation when the path length is one decimeter and the concentration of the sample in solution is one gram per milliliter. It is a measure of the degree to which a compound is dextrorotary or levorotary. Dextrorotary compounds have a positive specific rotation while levorotary compounds have negative. Mathematically it is defined as

$$[\alpha]_\lambda^T = \frac{\alpha}{l \times \rho}$$

for pure liquids. In this equation, α is the measured rotation in degrees for a sample at a temperature T and wavelength λ (in nanometers), l is the path length in decimeters, and ρ is the density of the liquid in g per mL.

For solutions, $[\alpha]_\lambda^T = \dfrac{\alpha}{l \times c}$

where c is the concentration of solution. Usually c is expressed in g/100mL and in that case a correction factor of 100 has to be included in the numerator,

$$[\alpha]_\lambda^T = \frac{100\,\alpha}{l \times c}$$

Although the formal unit of specific rotation is deg dm^{-1}cm^{3}g^{-1}, it is usually expressed as only degrees.

Q. 15. The specific rotation of a chiral compound is reported to be 10 deg dm^{-1}cm^{3}g^{-1}. What will be the expected observed rotation of 0.20M solution of the compound (molecular weight = 750 g mol^{-1}) in the same solvent at the same temperature?

Ans.: The specific rotation as defined in the previous question is given by the following expression.

$$[\alpha]_\lambda^T = \frac{100\,\alpha}{l \times c}$$

when concentration c is expressed in g/100mL.
Concentration of compound c = 0.20M = 0.20 mol L^{-1} = 0.20 x 750 =150 g L^{-1} = 15 g/100mL. $[\alpha]_\lambda^T$ = 10 deg dm^{-1}cm^{3}g^{-1}, l = 1 dm. Substituting the values in the above expression, the observed rotation, α = (10 x 1 x 15)/100 = 1.5 deg.

Q. 16. What is molecular rotation?

Ans.: Molecular rotation is a value obtained by multiplying specific rotation by the molecular weight of the substance expressed in g/mol. Molecular rotation is sometimes used in preference to specific rotation for comparing the optical rotations of different compounds on a molecular rather than weight basis. Its units are deg dm^{-1} cm^3 mol^{-1}.

Q. 17. What is the significance of the quantities specific rotation and molecular rotation?

Ans.: Specific rotation and molecular rotation are intrinsic properties for a particular substance at a given temperature and wavelength. They do not depend on the concentration of sample solution and other instrumental parameters such as length of the sample tube or cell. The values for many chiral compounds are available in the literature. The reported data can be used to quantify the chirality of a molecular species. The sign of the specific rotation value also tells us whether the compound is dextrorotatory or levo-rotatory. Moreover, the unknown concentration of the sample solution can be determined from the known literature value of specific rotation and measured optical rotation of solution.

Q. 18. The experimentally observed specific rotation ($[\alpha]_D^{25}$) for a sample was reported to be + 3.7⁰. What does it mean?

Ans.: The notation, $[\alpha]_D^{25}$ means that the measurement was conducted at 25⁰C using the D-line of the sodium lamp (λ=589.3 nm). A value of + 3.7⁰ means that a sample containing 1.00 g/mL of the compound in a 1 dm tube exhibits an optical rotation of 3.7⁰ in clockwise direction.

Q. 19. Explain how polarimetry can be used to determine the optical purity of enantiomers?

Ans.: Percent purity of enantiomer = $\dfrac{Observed\ specific\ rotation}{Specific\ rotation\ of\ pure\ enantiomer} \times 100$

For example, if the observed specific optical rotation of compound A is + 5.0⁰ while that for the corresponding pure enantiomer is + 25.0⁰, the percentage optical purity of compound is (+5.0/+25.0) x 100 = 20%. This means that the sample consists of 80 % of the racemic form (= equimolar mixture of both enantiomers for which α = 0⁰) and an excess of 20 % of the enantiomer in question. So the compound contains 80/2 (= 40%) + 20% = 60% of one enantiomer and 40% of the other enantiomer.

Q. 20. How this technique can be used to study the kinetics of hydrolysis of sucrose?

Ans.: Acidic hydrolysis of sucrose produces equivalent amounts of glucose and fructose.

Sucrose + Water → Glucose + Fructose

$$C_{12}H_{22}O_{11}(s) + H_2O~(l) \xrightarrow{H^+} C_6H_{12}O_6~(aq) + C_6H_{12}O_6~(aq)$$
   Sucrose   Water   Glucose   Fructose

Sucrose is dextrorotatory but after hydrolysis the solution becomes levorotatory because although glucose and fructose are produced in equivalent amounts, fructose is strongly levorotatory. That is the levorotatory power of fructose is greater than the dextrorotatory power of glucose at equal molar concentrations. Therefore, the optical rotation decreases from a positive value to zero and ultimately becomes negative as the reaction proceeds. On completion of reaction the reaction mixture is levorotatory. Due to this reason this reaction is also called inversion of sucrose. The change in the optical rotation of sucrose as a function of time can be used to study the kinetics of the reaction. The reaction is a pseudo first order reaction since one of the reactants is solvent water which is present in large excess and therefore, its concentration does not affect the rate of the reaction. The rate of the reaction, $dx/dt = k~[H_2O]~[sucrose]$. If the constant concentration of water is included in the rate constant and $k~[H_2O]$ is equal to another constant k_1 then rate, $dx/dt = k_1~[sucrose]$. If 'a' is the initial concentration of sucrose and 'x' is the amount hydrolyzed, then $(a - x)$ is the amount of sucrose remaining after time t. Therefore, $dx/dt = k_1~(a - x)$. By rearranging and integrating, we get $k_1 t = \ln~(a/(a - x))$. For an optically active substance, the amount of optical rotation is directly proportional to concentration. If α_0 is the initial angle of rotation of the mixture at zero time, α_∞ is the final angle of rotation when all sucrose has been inverted that is after completion of the reaction and α_t is the angle of rotation at time t, then 'a' is proportional to $(\alpha_0 - \alpha_\infty)$, 'x' is proportional to $(\alpha_0 - \alpha_t)$ and $(a - x)$ is proportional to $(\alpha_t - \alpha_\infty)$. Therefore, $k_1 t = \ln~[(\alpha_0 - \alpha_\infty)/(\alpha_t - \alpha_\infty)]$. A plot of $\ln~[(\alpha_0 - \alpha_\infty)/(\alpha_t - \alpha_\infty)]$ against time 't' gives a straight line passing through origin. The slope of the straight line is equal to the rate constant, k_1.

23. Dielectric constant and Dipole moment

Q. 1. What is a dielectric?

Ans.: A dielectric material or dielectric is an insulator which can be polarized by an applied electric field. They are frequently used in capacitors, radios and transmission lines for radiofrequency.

Q. 2. What is the difference between a dielectric and an insulator?

Ans.: Dielectrics are also type of insulators but they have the ability to get polarized on the application of an external electric field. Application of electric field shifts the electric charges from their average equilibrium positions causing dielectric polarization. The displacement of positive charges in the direction of the field and negative charges in the opposite direction results in the creation of an internal electric field. Porcelain (ceramic), mica and plastics are some examples of solid dielectric materials. Distilled water, hydrocarbon oils, silicone oils are the most common liquid dielectric materials while dry air, nitrogen and helium are excellent gaseous dielectrics. Perfect vacuum is also an exceptionally efficient dielectric.

Insulators are materials which do not allow the flow of electric current through them. Electrons in insulators are so closely and tightly joined to atoms by ionic or covalent bonds that they offer high resistance to the passage of electricity through them and therefore, no current flows through insulators even on the application of an external electric field. Fiberglass, silica, rubber are some examples of insulators. They differ in the field of application as well. Dielectrics are used to store the electric charges, while insulators are used to block the flow of electric charges.

Q. 3. Define dielectric constant.

Ans.: Dielectric constant, also called relative permittivity, measures the ability of a substance to become polarized when subjected to an electric field. In other words, it characterizes the tendency of the atomic charge in a dielectric material to distort in the presence of an electric field. The dielectric constant is defined as the ratio of the permittivity of a substance to the permittivity of free space (a vacuum). Electric permittivity is a constant of proportionality that relates the electric displacement and electric field intensity. This constant is equal to approximately 8.85×10^{-12} farad per meter (F/m) in vacuum. If ε_0 represents vacuum permittivity and ε represents the permittivity of the given substance, then the relative permittivity, also called the dielectric constant κ, is given by $\kappa = \varepsilon/\varepsilon_0$. Dielectric constant is always greater than or equal to 1. Dielectric constant cannot be less than one because the permittivity of the substance (ε) is always greater than the vacuum permittivity (ε_0).

Q. 4. What is the difference between permittivity and permeability?

Ans.: Permittivity and permeability are two commonly used terms in electromagnetism. Permittivity, which is related to the polarization of the material, measures the ability to form an electric field inside a medium in response to an external electric field. In other words, it characterizes the tendency of the atomic charge in a dielectric material to distort in the presence of an electric field. The larger the tendency for charge distortion, that is electric polarization, the larger the value of permittivity. On the other hand, the permeability of a material is related to the magnetization of the material and is a measure of the ability of the material to support the formation of a magnetic field within the material in response to an external magnetic field.

Q. 5. What is the unit for the dielectric constant?

Ans.: The Dielectric Constant of a medium is a number without units because it is a ratio.

Q. 6. What is the significance of dielectric constant in chemistry?

Ans.: Dielectric constant in general, measures the ability of a substance to insulate charges from each other. In chemistry, the dielectric constant of the solvent is an important parameter. The higher the dielectric constant of a solvent, the more polar it is and has greater ability to stabilize charges and hence dissolve salts. For example, a given solute dissociates into ions to a greater extent in water than in methanol since the dielectric constant of water (80) is higher than that of methanol (30). The dielectric constant of a solvent also influences the interactions in solution that involve ions and polar molecules.

Q. 7. Arrange the following organic solvents in the order of increasing dielectric constants: ethanol, acetonitrile, pyridine, chloroform, carbon tetrachloride, water.

Ans.: The polarity of the solvent can be a major guide to the numerical value of the dielectric constant of the solvent. Since dielectric constant increases with increase in the polarity of solvent, the polarity of the solvents and hence the dielectric constant values increase in the order carbon tetrachloride (2.2) < chloroform (4.8) < pyridine (12.4) < ethanol (24.5) < acetonitrile (37.5) < water (80.1).

Q. 8. What is the effect of temperature on dielectric constant?

Ans.: In general, dielectric constant decreases with increase in temperature. This is due to the effect of heat on orientation polarization. Dielectric materials have permanent dipoles. As temperature increases, the molecules in the dielectric have more thermal energy and therefore, the amplitude of random motion is greater. This means that at higher temperatures the dipoles are less aligned in the direction of the field resulting in reduced orientation polarization which is responsible for the decrease in the dielectric constant.

Q. 9. What is the dielectric constant of a metallic conductor?

Ans.: In metallic conductors, the mobile charge carriers are electrons which are free to move within the metal. Inside the conductor in equilibrium state, the charge carriers (electrons) are distributed uniformly in such a way that the net force on the electrons is zero and the electric fields cancel each other. In a metallic conductor charge resides only on the outer surface. Since the electric field inside a conductor is always zero under the static situation, the dielectric constant which is inversely proportional to electric field is infinite.

Q. 10. What is an electric dipole?

Ans.: A system consisting of two equal and opposite electric charges separated by a distance is called an electric dipole. In chemistry a dipole usually refers to the separation of charges within a molecule. In this context there can be three types of dipoles. Permanent dipoles occur when two atoms in a molecule have very different electronegativity, resulting in asymmetrical charge distribution. The more electronegative atom attracts electron pair towards itself thereby acquiring negative polarity while the other atom acquires positive polarity. Molecules with permanent dipoles are called polar molecules. The positive end of one molecule is aligned towards negative end of another molecule and so on. Dipole-dipole interactions occur between the molecules (intermolecular bonding) unlike covalent bonds which are between atoms in a molecule (intramolecular bonding). For example, hydrogen fluoride is a polar molecule since fluorine is more electronegative than hydrogen. Intermolecular forces are weak compared to intramolecular forces.

During random motion of electrons around nucleus, at any instant the electron distribution may be uneven or unsymmetrical resulting in creation of instantaneous or temporary dipoles. The instantaneous dipole-induced dipole forces, also called London dispersion forces, are smaller in magnitude than permanent dipole - dipole interactions. These weak intermolecular interactions are part of Van der Waal's forces and play an important role in the chemistry of non-polar molecules. For example, London dispersion forces cause non-polar molecules to condense to liquids and freeze into solids when temperature is lowered sufficiently. London dispersion forces are stronger between molecules that are easily polarized. Larger and heavier atoms and molecules also exhibit stronger dispersion forces than smaller and lighter ones.

Ion-dipole forces constitute the third type of intermolecular force which involves an electrostatic interaction between partially charged dipole of one molecule and a fully charged ion. The strength of the ion-dipole force is proportional to the ion charge. An ion-induced dipole interaction can also occur between a fully charged ion and a temporary dipole induced by the presence of ion. Ion-dipole forces are much weaker than covalent or ionic bonds but stronger than dipole-dipole interactions because the full charge of any ion is much greater than the charge of dipole.

Q. 11. What is dipole moment?

Ans.: Dipole moment is defined as the product of the magnitude of charge on the ends of dipole and the distance of separation between them. Dipole moment is a vector quantity and by convention its direction is from negative charge to positive charge. The molecules with a net dipole moment are polar molecules. It is a measure of the overall polarity of the charged system.

Q. 12. What are the units of dipole moment?

Ans.: Dipole moment is measured in the SI units of coulomb meter (Cm). However, since the charges in the molecules as well as the distance between them is very small, the magnitude of dipole moments is very small and therefore, another historical unit, the Debye (D) is also often used. 1 Debye = 3.33×10^{-30} Cm. Many molecules have dipole moments around 1 D.

Q. 13. Can dipole moment be zero? Give an example.

Ans.: Yes, dipole moment can be zero. The net dipole moment is not determined by the polarity of individual bonds but is dictated by the geometry of the molecule. The simplest example of a nonpolar molecule that contains polar bonds is carbon dioxide. This is a linear molecule with two polar C=O bonds. The central carbon will have a net positive charge and the two outer oxygen atoms a net negative charge. However, since the molecule is linear, these two bond dipoles cancel each other out (i.e. the vector addition of the dipoles equals zero) and the overall molecule has a zero dipole moment ($\mu = 0$).

Q. 14. Distinguish between polar and non-polar molecules.

Ans.: The polarity of a chemical bond is due to the unequal sharing of electrons between atoms forming the bond because of the electronegativity difference. Although a polar bond is necessary for a molecule to have a dipole, the over-all polarity of the molecule is determined by the geometry of the molecule. So a polar molecule has a net dipole moment while molecules with zero or very small dipole momenta are non-polar. A bond will only be slightly polar if the difference in the electronegativity of the atoms forming the bond is very small and the resulting molecules with very small overall dipole moments are also considered as non-polar. The most important example of this is hydrocarbons, the molecules that contain hydrogen and carbon. Since the electronegativity difference between H and C atoms is very small, C-H bond is very weakly polar and therefore, hydrocarbons with tiny dipole moments are many times considered as non-polar.

Q. 15. Define electric polarization and polarizability?

Ans.: The separation of the centre of positive and negative charges in a material caused by the application of an electric field is called electric polarization. The ease with which the electric distribution around an atom or molecule can be distorted resulting in separation of charges is defined as the polarizability of the material. As a result of the distortion of electron cloud, a non-polar molecule can also acquire induced dipole moment. This induced dipole moment (μ_{ind}) is related to the polarizability (α) of the molecule or atom and the strength of electric field E by the equation, $\mu_{ind} = \alpha E$. The polarizability α has units of Cm^2V^{-1}.

Q. 16. Define Electronic Polarization and distortion polarization.

Ans.: Electric polarization, defined as the displacement of positively charged nucleus and negatively charged electrons in opposite directions by an external electric field, is also called electronic polarization. For molecular substances, application of external electric field may also cause distortion of relative positions of nuclei in the molecule resulting in bending and stretching of the molecule. Distortion polarization is the sum of the contributions of electronic polarization and polarization due to the distortion of the molecular skeleton.

Q. 17. What is ionic polarization?

Ans.: Ionic polarization occurs in ionic crystals such as NaCl, KCl and LiBr. In the absence of an external electric field, there is no net polarization inside these materials. However, in the presence of an applied electric field, the displacement of cations and anions in opposite directions leads to induced polarization, called ionic polarization. The average induced dipole moment per ion pair (μ_{av}) is equal to the product of the ionic polarizability (α_i) and the local electric field (E) experienced by the ions ($\mu_{av} = \alpha_i$ E). Usually ionic polarizability is much higher than the electronic polarizability and therefore, ionic substances have high dielectric constants.

Q. 18. Define orientation polarization.

Ans.: Orientation polymerization occurs in molecules such as HCl and H_2O which have permanent dipole moment. In the absence of an electric field, the dipole moment is cancelled out due to thermal agitation. However, when an electric field is applied, the molecules rotate to align with the field causing a net average dipole moment per molecule. This preferential orientation of the dipoles in the field direction leads to orientation polarization. Unlike electronic and ionic polarization the orientation polarization is temperature-dependent and this is an important factor to be considered while selecting a dielectric material for electronic and optical applications.

Q. 19. Define Dielectric Loss.

Ans.: In an alternating current circuit, the periodic reversal of electrical charge leads to a loss of some amount of energy in the form of heat. This loss of energy is called dielectric loss. A good dielectric should have low dielectric loss. The heat produced through dielectric loss is also utilized for some industrial processes. For example, dielectric loss is utilized for heating food in a microwave oven since most of the dielectric loss is within the microwave range of electromagnetic radiation.

Q. 20. What is dielectric strength?

Ans.: Dielectric strength is defined as the maximum voltage or electric field that the material can withstand under ideal conditions without breaking down, that is, without losing its insulating capability. Above breakdown voltage, the electric field frees some bound electrons and the insulating material allows flow of charge that is, becomes electrically conducting. It is usually expressed in volts or kilovolts per unit thickness of material. Higher the dielectric strength of a material, the better is its quality as a dielectric. A perfect vacuum has highest dielectric strength because it does not contain any material which can break down.

Q. 21. What are Ferroelectric materials?

Ans.: Ferroelectric materials exhibit spontaneous electric polarization, even in the absence of electric field. Some common examples of ferroelectric materials are barium titanate ($BaTiO_3$) and Rochelle salt (sodium potassium tartrate tetrahydrate, $KNaC_4H_4O_6.4H_2O$). In such materials, the separation of the center of positive and negative charge makes one side of the crystal positive and the other negative. The direction of polarization can be reversed by inverting the direction of the applied electric field. Although the prefix ferro- means iron, most ferromagnetic materials do not contain iron. They show some similarities to ferromagnetic materials which exhibit permanent magnetic moment. All ferroelectric materials show piezoelectric effect. Piezoelectric effect refers to the ability of certain materials to generate an electric charge in response to an applied mechanical stress.

Q. 22. Is vacuum a Dielectric?

Ans.: Strictly speaking vacuum cannot be considered as a dielectric. A dielectric is defined as an insulating material which can be polarized by applying electric field. There is nothing which can be polarized in true vacuum. But in theory vacuum is usually considered as a dielectric medium of dielectric constant unity because it does not have any free charge carriers.

Q. 23. Is air a dielectric?

Ans.: Yes, dry air is an excellent dielectric. Air is an insulator but when very high electric field is passed through air, molecules in air get polarized. The relative permittivity of air changes with temperature, humidity and atmospheric pressure. At STP air has a relative permittivity of 1.0006. Air is the most common gaseous dielectric due to its easy availability and low cost. It is used in variable capacitors and some types of transmission lines.

Q. 24. What are the factors on which the dielectric constant of a material depends?

Ans.: The observed dielectric constant of a material depends on the following factors. i) Permanent dipole moment of molecules, ii) Polarizability of molecules in the applied electric field, iii) Density state of the material or concentration of related material when used as a solution between the capacitor plates and iv) Temperature.

Q. 25. Why are dielectrics used between the plates of a capacitor?

Ans.: A capacitor is a devise used to store electric charge. The capacitance is defined as the amount of charge stored in capacitor per volt. The capacitance (C) of a simple parallel plate capacitor is given by the equation $C = \varepsilon_0 (A/d)$. The constant ε_0 is the permittivity of free space; its numerical value in SI units is 8.85×10^{-12} Farad/m. A is the area of one plate in square meters and d is the distance between plates in meters. This equation is valid when the parallel plates are separated by air or free space. Thus for a small size capacitor the capacitance or the charge stored in capacitor is very small when the plates are separated by air or free space. However, when a dielectric with dielectric constant κ is placed between the plates of the capacitor, the capacitance is given by equation, $C = \kappa \varepsilon_0 (A/d)$ and therefore, capacitance increases by a factor of κ. High permittivity of the dielectric allows a greater stored charge at a given voltage. Thus the main purpose of the dielectric is to be able to store as much charge as possible in the capacitor. Moreover, dielectric material prevents the conducting plates, on which charges are stored in a capacitor, from coming into direct electrical contact and the distance d between the plates can be made as small as possible which further increases capacitance. Dielectric strength, defined as the electric field strength above which the material begins to break down and conduct, imposes a limit on the voltage that can be applied for a given plate separation.

Q. 26. How dipole moment is determined experimentally?

Ans.: The most common approach is to calculate dipole moment from the dielectric constant of materials. The permanent dipole moment of a polar solute molecule can be determined experimentally from measurement of the dielectric constant (κ) of solution of the polar solute in a nonpolar solvent.

Q. 27. How dielectric constant is measured experimentally?

Ans.: The dielectric constant is measured as the ratio of the capacitance of a parallel plate capacitor with the dielectric (test material) between the plates to the capacitance of the same capacitor with vacuum (or air) between them. If C is the capacitance of the cell when the medium between the condenser plates is the solution of the test substance, and C_0 is the capacitance when the medium is vacuum then dielectric constant, $\kappa = C/C_0$. Since air has a dielectric constant very close to that of vacuum, C_{air} is usually used in place of C_0 in the above expression.

Q. 28. Do materials with molecules of higher dipole moments will have greater dielectric constants also?

Ans.: This is only qualitatively true. The dielectric constant and dipole moment of a substance do not have a direct proportionality. This is due to distortion or polarization produced in molecules by the applied electric field.

Q. 29. How experimentally determined dielectric constant data can be used to determine the molar polarization and dipole moment?

Ans.: The molar polarization (P_M) can be determined from the experimentally determined dielectric constant (κ) using Clausius-Mossoti equation
$$\frac{\kappa - 1}{\kappa + 2}(M/\rho) = \frac{1}{3\varepsilon_0}N_A\alpha = P_M$$
In this equation, M is the molar mass and ρ is the density of the dielectric material, ε_0 is the vacuum permittivity and α is the polarizability. If the molecules have no permanent dipole moment, only distortion polarization takes place. The corresponding polarizability is denoted by α_0. If each molecule has a permanent dipole moment of magnitude, μ there is a tendency for the moment to become oriented parallel to the field direction and the resulting contribution to polarizability is given by $\mu^2/3kT$ where k is the Boltzmann constant. Thus the total polarizability is given by
$$\alpha = \alpha_0 + \mu^2/3kT$$
and molar polarization can be written in the form
$$P_M = \frac{\kappa - 1}{\kappa + 2}(M/\rho) = \frac{1}{3\varepsilon_0}N_A(\alpha_0 + \mu^2/3kT) = P_d + P_\mu$$

P_d and P_μ are the distortion and orientation contributions to the molar polarization, respectively. This is the equation of a straight line and so both α_0 and μ can be obtained from the intercept and slope of P_M versus $1/T$ plot.

Q. 30. What are the units of molar polarization?

Ans.: The dielectric constant is unit less and so according to Clausius-Mossoti equation, defined in Q. 29, molar polarization has units of M/ρ, Molar mass (kg mol^{-1})/Density (kg m^{-3}) = m^3/mol. Thus molar polarization has same units as molar volume (m^3/mol).

Q. 31. How dipole moment of a polar solute in a nonpolar solvent can be determined experimentally.

Ans.: For measurements in solution, if we consider a dilute solution of a polar solute 2 in a nonpolar solvent 1, the molar polarization can be written as

$$P_M = x_1 P_{1M} + x_2 P_{2M} = \frac{\kappa - 1}{\kappa + 2} \frac{(M_1 x_1 + M_2 x_2)}{\rho} \quad (1)$$

Where x's are the mole fractions, M's are molecular weights, κ is the dielectric constant and ρ is the density of the solution. The subscripts 1 and 2 refer to the solvent and solute, respectively. The molar polarization of the pure solvent, P_{1M}, can be assumed to have the same value in solution as in the pure solvent because the nonpolar solvent has only distortion polarization, which is not greatly affected by interactions between molecules. Therefore,

$$P_{1M} = \frac{\kappa_1 - 1}{\kappa_1 + 2}(M_1/\rho_1) \quad (2)$$

Rearrangement of equation (1) leads to

$$P_{2M} = \frac{1}{x_2}(P_M - x_1 P_{1M}) \quad (3)$$

Due to strong interactions between permanent dipoles of solute molecules, the molar polarization of solute in solution (P_{2M}) generally decreases with increase in the mole fraction of solute, x_2. We know that in an infinitely dilute solution since the solute molecules are far apart, the interactions between permanent dipoles of solute are minimized and the solution approaches ideal behavior. The molar polarization of solute at infinite dilution (P^0_{2M}) cannot be obtained by extrapolation of P_{2M} versus x_2 plot to $x_2 = 0$ since the plot is not linear. This is usually done as follows. If we assume linear dependence of the dielectric constant (κ) and the density of solution (ρ) on the mole fraction of solute in solution (x_2) then we can write

$$\kappa = \kappa_1 + ax_2 \quad (4)$$
$$\rho = \rho_1 + bx_2 \quad (5)$$

where a and b are constants. On substituting equations (2), (4) and (5) in equation (3) and rearranging terms we obtain the limiting expression

$$P_{2M}^0 = \frac{3M_1 a}{(\kappa_1 + 2)^2 \rho_1} + \frac{\kappa_1 - 1}{(\kappa_1 + 2)\rho_1}\left(M_2 - \frac{M_1 b}{\rho_1}\right) \quad (6)$$

The values of constants 'a' and 'b' are obtained from the slopes of the linear κ versus x_2 and ρ versus x_2 plots, respectively. Thus limiting molar polarization of the solute in solution can be calculated from equation (6).

If we assume that the molar distortion polarization in an infinitely dilute solution is equal to that of the pure solvent, then

$$P_{2d}^0 = \frac{n_2^2 - 1}{n_2^2 + 2} \quad (7)$$

Where n_2 is the refractive index and ρ_2 is the density measured for the solute in the pure state. The molar orientation polarization of the solute at infinite dilution ($P^0_{2\mu}$) can be obtained as the difference between the P^0_{2M} values calculated from equation (6) and P^0_{2d} values calculated from equation (7). It has been shown in Q. 29 that

$$P_M = \frac{1}{3\varepsilon_0} N_A \left(\alpha_0 + \mu^2/3kT\right) = P_d + P_\mu$$

So we can write

$$P_{2\mu}^0 = P_{2M}^0 - P_{2d}^0 = \frac{1}{3\varepsilon_0} N_A \left(\mu^2/3kT\right) \quad (8)$$

On substituting the numerical values of physical constants, we obtain

$$\mu = 42.7 \, (P_{2\mu}^0 T)^{1/2} \times 10^{-30} \, Cm = 12.8 \, (P_{2\mu}^0 T)^{1/2} \, Debye \quad (9)$$

where $P^0_{2\mu}$ is given in m^3 mol^{-1} units and T is in kelvin. (1 Debye, D = 3.33564 × 10^{-30} C.

General Bibliography

1. Shoemaker D. P., Garland C. W., Nibler J. W. Experiments in physical chemistry. 5th Edition, McGraw-Hill Book Company, Singapore, 1989.

2. Das R. C., Behera B. Experimental physical chemistry. Tata McGraw-Hill Publishing Company Limited, New Delhi, 1983.

3. Levitt B. P. Finlay's practical physical chemistry. 9th Edition, Longman, London and New York, 1973.

4. Khosla B. D., Garg V. C., Gulati A., Senior practical physical chemistry. 16th Edition, R Chand & Co., New Delhi.

5. CRC Handbook of Chemistry and Physics. Editor-in-Chief John R. Rumble, 98th edition, CRC Press, 2017.

6. Wilson J. M., Newcombe R. J., Denaro A. R., Rickett R. M. W., Experiments in physical chemistry. Second edition, Pergamon Press, 2016.

7. Viswanathan B., Raghavan P. S., Practical Physical Chemistry. Viva Books, 2014.

8. Kaye G. W. C., Laby T. H., Handbook of physical and chemical constants. 16th Edition, Longmans, Green and Co., London, 1995.

9. Young H. D. Statistical treatment of experimental data. McGraw-Hill. 1962.

10. Taylor J. R. An introduction to error analysis: The study of uncertainties in physical measurements. University Science Books, 2nd edition, 1996.

11. Gurtu J. N., Gurtu A., Advanced Physical chemistry experiments. Pragati Prakashan, Meerut, 2000.

12. Weissberger A., Rossiter B. W. (eds.), Techniques of chemistry, Vol. I Physical Methods of Chemistry, Wiley-Interscience, New York. Part I (1971) – part VI (1977).

13. Rossiter B. W., Hamilton J. F. (eds.), Physical methods of chemistry, 2nd Edition, Vol. I and II, Wiley-Interscience, New York, 1986.

14. Weast R. C. (ed.), Handbook of chemistry and physics, 69th edition, CRC PressBoca Raton, Fla. 1988/89.

15. Journal of Physical and Chemical Reference Data, Published by American Chemical Society and American Institute of Physics, Washington D. C., regular issues since 1972 plus any supplements.

16. Maity S. K. and Ghosh N. K., Physical Chemistry Practical, New Central Book Agency, 1st Ed. 2012.

17. Athawale V. D. and Mathur P., Experimental Physical Chemistry, New Age International Publishers, 1st Ed. 2001.

18. Halpern A. and McBane G. Experimental Physical Chemistry: A Laboratory Text, W. H. Freeman, 3rd Ed. New York, 2006.

19. Yadav J. B., Advanced Experimental Physical Chemistry, Krishna Prakashan Media (P) Ltd. 2015.

20. Thorat B. R. Handbook of Experimental Physical Chemistry, Scholars' Press, 2019.

www.ingramcontent.com/pod-product-compliance
Lightning Source LLC
Chambersburg PA
CBHW060824220526
45466CB00003B/961